SUBSTANCE ABUSE ASSESSMENT, INTERVENTIONS AND TREATMENT

NICOTINE DEPENDENCE, SMOKING CESSATION AND EFFECTS OF SECONDHAND SMOKE

SUBSTANCE ABUSE ASSESSMENT, INTERVENTIONS AND TREATMENT

Additional books in this series can be found on Nova's website under the Series tab.

Additional e-books in this series can be found on Nova's website under the e-book tab.

SUBSTANCE ABUSE ASSESSMENT, INTERVENTIONS AND TREATMENT

NICOTINE DEPENDENCE, SMOKING CESSATION AND EFFECTS OF SECONDHAND SMOKE

CARLTON L. OWENS
EDITOR

nova publishers
New York

Library of Congress Cataloging-in-Publication Data

Nicotine dependence, smoking cessation and effects of secondhand smoke / editor, Carlton L. Owens.
pages cm. -- (Substance abuse assessment, interventions and treatment)
Includes index.
ISBN 978-1-63482-642-6 (hardcover)
1. Nicotine addiction. 2. Tobacco use--Health aspects. 3. Smoking cessation. 4. Passive smoking--Health aspects. I. Owens, Carlton L.
RC567.N5236 2015
616.86'5--dc23

2015012839

Published by Nova Science Publishers, Inc. † New York

CONTENTS

PREFACE

The clinical course of nicotine dependence, nicotine withdrawal and smoking cessation varies significantly between men and women smokers. Recent research has also suggested that certain personality traits and experiences predict and help maintain smoking behavior, thus leading to dependence and negative health consequences. The chapters in this book include research on the correlation between people's early engagement in smoking behavior and later problematic levels of use, the role of hormone therapy (HT) to treat smoking/nicotine dependence; the association between active and passive water pipe tobacco smoking and health issues that might develop later on (as well as the similarities and differences between tobacco smoking and cigarette smoking).

Worldwide tobacco use, production and consumption are continuing. Of approximately 1.2 billion smokers in the world, about half of them will die from diseases caused by smoking. The authors' describe the extent of nicotine dependence, the neurobiology of nicotine dependence, which is important for understanding tobacco addiction and its treatment, and describe the pharmacological and non-pharmacological treatments of nicotine dependence, including electronic cigarettes.

The final two chapters detail the chemical composition and health risks associated with secondhand smoke (SHS) exposure. These chapters address the impact of SHS on children in particular, as well as the deleterious effect on workers' health that often takes place in indoor environments, and finally, recent global public policies that have been implemented on tobacco control are addressed.

Chapter 1 – The clinical course of nicotine dependence, nicotine withdrawal, and smoking cessation varies significantly between women and men smokers. Hormonal differences between the two sexes primarily account for this variation in pharmacokinetics and pharmacodynamics. Although these intersex differences are highly important and interesting, this chapter will have an intrasex focus on women and the role of hormone therapy (HT) in the clinical course of smoking/nicotine dependence. Most empirical studies in this area have examined HT's influence on smoking cessation outcomes and nicotine withdrawal symptoms, including mood, appetite, and weight changes. Collectively, the existing studies show no evidence to suggest that HT significantly influences smoking cessation outcomes, nicotine withdrawal symptoms, postcessation appetite change, or weight gain among women smokers. However, studies investigating the effects of HT on postcessation mood and affect have yielded mixed results. Recent research highlighting the role of monoamine oxidase inhibitor (MAOI) content in cigarette smoke and its antidepressant effects, as well as documented increases in MAO with the declining estrogen levels that define the menopausal transition are incorporated to help draw conclusions. There are several important considerations in attempting to draw meaningful conclusions from these studies as well as in moving forward with new research in this area; these include: (1) the use of pharmacological intervention, such as bupropion or nicotine replacement therapy as smoking cessation aids in clinical trials and how they interact with HT to influence withdrawal symptoms and cessation outcomes; (2) the self-selection for HT use versus random assignment to receive HT among women participating in existing studies; and (3) the difficult distinction between mood, appetite/weight, and other physiological symptoms related to nicotine dependence or withdrawal, and those symptoms related to the menopausal transition. In the present chapter, we will provide a comprehensive review of the existing studies addressing HT's role in nicotine dependence and smoking cessation among women, highlight consistent themes and limitations in the existing literature, and finally, identify and offer recommendations for future areas of investigation.

Chapter 2 – Recent research has suggested that certain personality traits and experiences predict and help maintain smoking behavior, thus leading to dependence and negative health consequences. A crucial challenge is to understand how people's early engagement in smoking behavior develops into problematic levels of use, despite their knowledge of the harms associated with smoking. In this chapter, we describe evidence indicating that there appears to be a positive feedback loop of increasing smoking and increasing risk for future smoking due to transactional processes involving high-risk personality traits, high-risk psychosocial learning, and early adolescent smoking behavior. A striking aspect of the findings supporting this model is that there is no evidence of a corrective process, whereby early smoking experiences lead to reduced risk. The opposite appears to be the case: early smoking leads to greater personality-based and learning-based risk. Remarkably, this process occurs in a climate in which virtually everyone knows about the health risks of smoking.

Chapter 3 – Approximately 100 million people smoke water pipe (shisha, hookah, narghile) every day. This trend is increasing in the Middle East, US and in Europe especially in adolescents. The reason for the widely distributed use of water pipe smoking is nurtured by the misleading assumption that this form of smoking is less harmful than cigarette smoking. Studies revealed that tobacco smoke from water pipe, depending on the used tobacco and charcoal, contains higher concentrations of nicotine, carbon monoxide, carboxy hemoglobin, heavy metals, carcinogens and fine dust particles compared to cigarette smoke. Similar to cigarette smoking, water pipe tobacco smoking causes a severe health risk by second hand smoke, because the side-stream smoke of the water pipe contains higher concentrations of polycyclic aromatic hydrocarbons, aldehydes and carbon monoxide compared to the side-stream smoke of cigarettes. According to studies, active and passive water pipe tobacco smoking was significantly associated with lung cancer, chronic obstructive lung diseases, asthma, cardiovascular effects, periodontal diseases and low birth-weight.

Chapter 4 – Worldwide tobacco use, production, and consumption are continuing. Of approximately 1.2 billion smokers in the world, about half of them will die from diseases caused by smoking. Tobacco use is a major avoidable cause many diseases. Despite of recommendations to stop smoking, smokers with life-threatening medical disorders are unable to quit smoking. Approximately, 80% of American smokers who made attempt to quit smoking independently by themselves, relapsed within the first month of abstinence. These quitting attempt failures may relate to nicotine dependence and/or low self-efficacy to quit. In 1988, the Surgeon General's report on the health consequences of smoking definitively concluded that all tobacco forms were powerfully "addicting drugs." Nicotine is a pharmacologic agent that has physiological and psychological effects on the human body. Nicotine has effects on mood and cognitive function that contribute to tobacco addiction. Smokers report positive effects such as pleasure and relaxation, which triggers them to smoke.

According to treating tobacco use and nicotine dependence guidelines, tobacco dependence is recognized as a chronic disease and requires serious interventions. Five nicotine replacement therapies (NRTs: nicotine inhalers, lozenges, gum, nasal sprays, and patches) and two non-NRTs (bupropion SR, and varenicline) are the first line to treat nicotine dependence. Nicotine replacement medications are recommended to maintain abstinence from tobacco, avoid nicotine withdrawal symptoms and prevent relapse. Combinations of counseling and medication are recommended to achieve effective treatment and increase long-term smoking cessation rates.

The goal of this review chapter is to describe the nicotine dependence, discuss the neurobiology of nicotine dependence, which is important for understanding tobacco addiction and its treatment, and to describe the pharmacological and non pharmacological treatments of nicotine dependence.

Chapter 5 – Tobacco smoke is a commonly encountered air pollutant linked to chronic respiratory diseases, cardiovascular disease, and cancer in non-smokers. Non-smokers are exposed to tobacco smoke through secondhand smoke (SHS) exposure. Children are the most vulnerable population who are afflicted by SHS. Nationally, more than 18% of 3 to 11 year olds live with a smoker. More than 50% of children in the U.S. are exposed to second hand smoke. Children between the ages of 3 to 11 years have the highest levels of SHS exposure (60.5%), likely because they spend a majority of their time in close proximity to a smoking caregiver. In epidemiological studies, SHS exposure decreases lung growth and increases the incidence of sudden infant death, respiratory illness, middle-ear disease, wheezing, and asthma. Children with chronic lung disease are a particularly vulnerable population who are significantly affected by SHS. For example, children with asthma and cystic fibrosis are negatively impacted by SHS. Asthma is the most common childhood chronic illness, affecting 9.6% of youth in the U.S. SHS exposure in children with asthma is related to increased frequency and severity of asthma and asthma exacerbations. Similarly, children with cystic fibrosis who are exposed to SHS have an increased number of pulmonary exacerbation and hospitalizations. This chapter will address the impact of SHS on children with chronic lung disease, and interventions health care workers can implement to decrease SHS exposure in these children.

Chapter 6 – Electronic cigarette is the most known electronic nicotine delivery device which was invented in 2003 and it is widely used by thousands of people nowadays. Its supporters consider it safe because it does not contain the harmful substances which are produced by the combustion of tobacco in conventional cigarettes. However, plethora of studies showed that the vapour of e-cig contains toxic and dangerous substances as the smoke of conventional cigarette but in different quantities. E-cig has been implicated for passive smoking and adverse effects on general and pediatric population. The use of e-cig is considered an alternative way of smoking. Its safety and its efficacy on smoking cessation have not been proved yet. More and longer lasting studies are needed.

Chapter 7 – Secondhand smoke is the result of the stream produced by the burning cone in every puff, the exhaled smoke and the air that is in the room. The study of chemical composition of secondhand smoke has shown that is a complex mixture of constituents and the concentration of every compound depend on time and environmental conditions. There are different respirable suspended particles in secondhand smoke like nicotine, carbon monoxide, benzene and formaldehyde, among others. Nicotine in hair is the most used biomarker in population-based studies.

Exposure to secondhand smoke takes place in indoor environments. Several studies in bar and restaurant's personnel have shown that there is a dose-response between indoor vapor-phase nicotine in bars or restaurants and the employee's hair nicotine concentration. This exposure has deleterious effects on the worker's health. Respiratory symptoms like wheeze, phlegm and throat pain have been described. Psychological effects are observed as well, including depressed mood, insomnia and anxiety. Furthermore, general population's health is also affected by secondhand smoke. In countries where a complete legislation ban to smoke in all indoor public places was implemented, a reduction in the prevalence rates of myocardial infarction after the implementation was reported.

According to the World Health Organization Framework Convention on Tobacco Control, countries that signed must implement comprehensive smoke-free legislation that prohibit tobacco smoking in all indoor public places, nevertheless, there are countries that still do not legislate, allowing to smoke in all areas, or have partial ban legislations, consenting to have mixed-areas inside a venue.

In: Nicotine Dependence
Editor: Carlton L. Owens
ISBN: 978-1-63482-642-6

Chapter 1

THE ROLE OF HORMONE THERAPY IN NICOTINE DEPENDENCE AND SMOKING CESSATION

Amy L. Copeland, Ph.D., M.P.[1]***, Krystal Waldo, M.A.***[1]***, MacKenzie R. Peltier, M.A.***[1] ***and Gerald S. Hecht***[2]
[1]Louisiana State University, Department of Psychology
Baton Rouge, LA
[2]Southern University
Department of Psychology
Baton Rouge, LA

ABSTRACT

The clinical course of nicotine dependence, nicotine withdrawal, and smoking cessation varies significantly between women and men smokers. Hormonal differences between the two sexes primarily account for this variation in pharmacokinetics and pharmacodynamics. Although these intersex differences are highly important and interesting, this chapter will have an intrasex focus on women and the role of hormone therapy (HT) in the clinical course of smoking/nicotine dependence. Most empirical studies in this area have examined HT's influence on smoking cessation outcomes and nicotine withdrawal symptoms, including mood, appetite, and weight changes (Allen et al., 2003; Allen et al., 2004; Oncken et al., 2007). Collectively, the existing studies show no evidence to suggest that HT significantly influences smoking cessation outcomes, nicotine withdrawal symptoms, postcessation appetite change, or weight gain among women smokers. However, studies investigating the effects of HT on postcessation mood and affect have yielded mixed results. Recent research highlighting the role of monoamine oxidase inhibitor (MAOI) content in cigarette smoke and its antidepressant effects, as well as documented increases in MAO with the declining estrogen levels that define the menopausal transition are incorporated to help draw conclusions. There are several important considerations in attempting to draw meaningful conclusions from these studies as well as in moving forward with new research in this area; these include: (1) the use of pharmacological intervention, such as bupropion or nicotine replacement therapy as smoking cessation aids in clinical trials and how they interact with HT to influence withdrawal symptoms and cessation outcomes; (2) the self-selection for HT use versus random assignment to receive HT among women participating in existing studies; and (3) the difficult distinction between mood, appetite/weight, and other physiological symptoms related to nicotine dependence or withdrawal, and those symptoms related to the menopausal transition. In the present chapter, we will provide a comprehensive review of the existing studies addressing HT's role in nicotine dependence and smoking cessation among women, highlight consistent themes and limitations in the existing literature, and finally, identify and offer recommendations for future areas of investigation.

Introduction: Menopausal Women and Smoking

Cigarette smoking has serious health consequences for individuals, including cardiovascular disease, stroke, respiratory illness, and cancer of the lungs, stomach, bladder and kidneys (Center for Disease Control; CDC, 2011). Women are at greater risk than men for experiencing certain smoking-related illnesses, such as lung cancer (Ben-Zaken Cohen, Pare, Man, & Sin, 2007; Zang & Wynder, 1996). Women smokers who have reached menopause (defined as occurring a year after a woman's final menses) are even more vulnerable than other women smokers to adverse health effects. This is largely because cigarette smoking has antiestrogenic properties and has been associated with altered estrogen metabolism and serum estrogen levels (Michnovicz, Hershcopf, Naganuma, Bradlow & Fishman, 1986; Tanko & Christensen, 2004).

It is necessary to consider the effects of HT on smoking cessation-related variables when discussing smoking and smoking cessation in menopausal women. The 2002 dissemination of the Women's Health Initiative (WHI) data showing an increased risk of breast cancer and heart disease in women randomized to HT led to a substantial reduction in its use (Barbaglia et al., 2009; Majumdar, Almasi, & Stafford, 2004). More recent data has further highlighted the potential for negative health consequences of HT (Chlebowski et al., 2010; Marshall et al., 2010). Furthermore, smokers appear to be at increased risk for HT-related health problems compared to nonsmokers (Wilson, Garrison, & Castelli, 1985; Chlebowski et al., 2010). For example, Chlebowski et al. (2010) found a greater increase in risk of lung cancer death from HT use in smokers compared to nonsmokers. However, the use of HT as a short term treatment for younger peri- and postmenopausal women appears to have fewer health risks (Rossouw et al., 2007), and HT continues to be prescribed by physicians. Therefore, a review of the influence of HT on smoking cessation is warranted based on both its continued use by women smokers, as well as theoretical interest in the effects of female sex hormones on smoking behavior.

The existing empirical studies in this area have examined HT's influence on smoking cessation outcomes and nicotine withdrawal symptoms, including mood, appetite, and weight changes (Allen et al., 2003; Allen et al., 2004; Oncken et al., 2007). In the following three sections, we review the literature relevant to each of these areas. Following a review of the pertinent empirical literature in each of these areas in which we highlight consistent themes and limitations in the existing literature, we will identify and offer recommendations for future areas of investigation.

HT and Mood in Smoking Cessation

Two studies were identified that investigated the effects of HT on mood and affect during smoking abstinence or cessation. This is an area of interest and importance given that smoking and negative affect or clinical depression occur comorbidly, and that depression that occurs postcessation can be a risk for relapse (e.g., Brandon, 1994). Allen et al. (2003) investigated whether transdermal hormone replacement therapy influenced smoking cessation variables, including depression, in postmenopausal smokers during short-term abstinence from cigarettes. Women currently on HT (*n*=17) and those not on HT (*n*=13) were recruited into respective groups. The HT group then had their medication replaced with a standard 0.1 mg estradiol transdermal system and 2.5 mg of Cycrin daily. After 3 weeks, baseline measures were taken, and participants were instructed to quit smoking for 2 weeks. They were provided with smoking-cessation counseling and monitored for abstinence. Depression was assessed with the Beck Depression Inventory (BDI), and BDI change scores (week 2 – baseline) differed significantly for the HT and non-HT groups, with women in the HT group experiencing an increase in depressive symptomatology. This finding was contrary to the investigator's hypothesis. They concluded that the findings may have clinical implications for postmenopausal women attempting to quit smoking while on HT, particularly since depressed mood is associated with a relapse to smoking.

In another study of postmenopausal women smokers (*N*=94) conducted by Allen et al. (2004), participants were grouped according to whether or not they were currently using HT and were then randomized within groups to active nicotine or placebo nicotine transdermal patch. Participants subsequently received cessation counseling and were monitored for abstinence during a 2-week period in which they quit smoking. They were assessed at baseline and during cessation on measures of acute symptomatology, including nicotine withdrawal, cigarette craving, smoking urges, mood changes, depressive symptoms, motor speed, and reaction time to determine if the beneficial effects of transdermal nicotine replacement therapy (NRT) differed for those who use and do not use HT. Two-way analyses of covariance (ANCOVAs) were run on change scores for dependent variables, with nicotine patch group (active/placebo) and HT group (HT/non-HT) as independent variables and age as a covariate. No interactions were found between HT and patch condition, but both showed specific effects. During the first week of abstinence, participants on active nicotine patch reported lower severity of withdrawal, greater reductions in cigarette cravings, and lower smoking urges, as compared to those on the placebo patch. During the second week of abstinence, participants using HT reported better mood on the Profile of Mood States (POMS) and less depression as per BDI scores, as compared to those not using HT. The authors concluded that the efficacy of transdermal nicotine replacement is not adversely modified by women's HT use, and that ovarian hormones might influence women's responses to smoking cessation, and thus should be considered in developing effective strategies for women to quit smoking.

In summary, two studies exist in which the effects of HT on cessation-related mood changes experienced by women smokers were examined. Both studies appear to have been conducted with methodological rigor, and both had a priori predictions regarding the benefit of HT use on mood during cessation. In both studies, the primary measure used to assess mood was the BDI, so there is not a concern of method variance in the study outcomes.

Rather, it seems likely that the variance lies in the samples of women smokers in the studies. It is possible that the samples were not homogenous in menopausal status. That is, whether the women were peri- versus postmenopausal, the former referring to the phase just prior to postmenopause. McVay and Copeland (2011) were careful to point out this distinction in a recent review of smoking cessation in peri- and postmenopausal women smokers. Although women may be using HT during either of these points in the menopausal transition, the former is associated with substantially fluctuating levels of estrogen. Indeed, at times estrogen levels during perimenopause are elevated above premenopause levels (Hale et al., 2007; Prior, 1998). Greater heterogeneity in one study sample versus another regarding inclusion of peri- and postmenopausal women is one possible explanation for the mixed findings regarding mood, cessation and HT.

HT and Post-cessation Appetite Changes

Despite popular belief, research does not support the idea that smoking cigarettes suppresses appetite and food intake in male and female smokers; however, there is evidence that food intake increases during a smoking cessation attempt among both genders (Perkins, 1992). Longitudinal studies have demonstrated that food intake increases, especially sucrose, fat and carbohydrate consumption, during the first few weeks following a cessation attempt (Perkins, 1992; Hall, Ginsbert, & Jones, 1986; Hatsukami, LaBounty, Hughes, & Laine, 1993). Preclinical studies have demonstrated that food intake and appetite increase as estrogen levels decrease in rodents (Wade, 1975). However, clinical study results do not replicate this relationship, suggesting that the estrogen deficiency observed in postmenopausal women may not affect appetite and food intake in the same way as observed in preclinical trials, especially following a cessation attempt (Duval, Prud'homme, Rabasa-Lhoret, Strychar, Brochu, Lavoie & Doucet, 2013).

For instance, Allen and colleagues (2005) studied the post-cessation nutritional (i.e., carbohydrate, sweet carbohydrate and fat) intake of 94 postmenopausal female smokers, who were either placed on hormone therapy (HT; daily regimen of 25 cm^2 transdermal estradiol and 2.5 mg oral medroxyprogesterone acetate) or no HT. No significant differences in carbohydrate, sweet carbohydrate or fat intake were observed between the HT users and the non-HT users after two weeks of smoking abstinence. These results indicate that the proposed appetite-depressive effect of estrogen does not significantly impact post-menopausal females, post-cessation attempt (Allen, Hatsukami, Britnell & Bade, 2005).

This conclusion is further supported by similar research in which no differences in nutritional (i.e., carbohydrate, sweet carbohydrate and fat) intake between post-menopausal women, who were stratified, based upon HT regimen (the same daily regimen described above versus non-HT users) after two weeks of smoking cessation (Allen, Brintnell, Hatsukami & Reich, 2004). It should be noted that the above mentioned studies only studied a single formulation/dose of HT, over a short period of time (two weeks of follow-up); thus additional studies are needed to determine the effects of HT on appetite in long-term quit attempts, as well as how different formulations and doses of HT affect appetite and food intake.

Despite little evidence that estrogen deficiency and HT impact appetite changes in post-menopausal women during a cessation attempt, there is evidence that nicotine replacement therapy (NRT) may increase appetite in this population. During a two week cessation attempt, post-menopausal research participants (both HT users and non-HT users) using a transdermal nicotine patch, demonstrated increased fat, carbohydrate and sweet carbohydrate intake, as well as increased total caloric intake, when compared to participants using a placebo patch during the quit attempt (Allen et al., 2005). Additionally, a similar study found post-menopausal women treated with an active transdermal nicotine patch reported more total calorie intake and higher sugar intake 12 weeks into a cessation attempt, than counterparts using a placebo transdermal patch (Allen et al., 2013). Conversely, there is evidence that those post-menopausal women randomized to abstain from smoking with no NRT, reported higher total caloric intake, as well as increased carbohydrate and sweet carbohydrate intake when compared to their counterparts who continued to smoke. There were no differences observed for fat consumption (Allen et al., 2004).

This indicates that nicotine consumption may have potential to attenuate appetitive intake of total calories, including carbohydrates and sweet carbohydrates across both HT and non-HT users who are making a cessation attempt. This conclusion is consistent with previous research that appetite and food intake increase following a cessation attempt and NRT has shown promising preliminary evidence to corroborate these findings (Perkins, 1992; Allen et al., 2005). However, it still remains unclear if NRT alleviates post-cessation appetitive changes beyond 3 months of abstinence in post-menopausal women, due to the limited evidence to date. Additionally, the clinical impact of this proposed alleviation is uncertain, as no clinically significant decreases in weight gain have been reported among studies of NRT-using post-menopausal women (Allen et al., 2013).

HT and Weight Change

Past research has shown that many postmenopausal women experience weight gain during the menopausal transition period. For example, in a large study of peri-menopausal women, Wing, Matthews, Kuller, Neilahn, and Plantinga (1991) found an average weight gain of 2.3 kg over a 3-year period with 20% gaining 4.5 kg or more. Given this weight gain related to hormonal changes, many older female smokers have major concerns regarding additional post-cessation weight gain. Since weight gain appears to be a major concern for older women, more research on smoking cessation interventions that address weight gain in older women is needed.

Increased fear of post-cessation weight gain due to both changes in metabolic rates and increases in food intake is an obstacle to smoking cessation, particularly for older women (Filozof, Fernandez, Pinilla, & Fernandez-Cruz, 2004; Hatsukami, LaBounty, Hughes, & Lane, 1993). Past research has shown that women tend to gain more weight than men (Nides et al., 1994; Hatsukami et al., 1995; Gritz, Nielsen, & Brooks, 1996) and thus are more likely to relapse as a result of post-cessation weight gain (Swan, Ward, Carmelli, & Jack, 1993). In a study examining weight gain following smoking cessation, O'Hara et al. (1998) reported that 19.1% of women who quit smoking gained greater than 20% of their body weight compared to 7.6% in men. Additionally, previous research indicates that women of greater age (e.g. 56-64 years old) gain more postcessation weight than younger premenopausal women (Kasteridis &Yen, 2012; Caan et al., 1996; Copeland et al., 2006) and that postmenopausal women get particular weight control benefits from smoking.

Given the combination of menopause and smoking cessation on weight gain in postmenopausal women, quitting smoking during this time may increase a women's risk for postcessation weight gain, and thus, may contribute to reduced motivation to quit smoking. Additionally, another study showed that postmenopausal women are more prone to weight gain following smoking cessation, regardless of outcome, with those who are successful in quitting gaining more weight (Korhonen, Goodwin, Miesmaa, Dupuis, & Kinnunen, 2011; Kleppingerm, Litt, Kenny, Oncken, 2010) and with the average weight gain approximately 10 pounds, six months after quitting

(Cropsey et al., 2010). Given the reduced benefit of quitting due to post-cessation related weight gain (Luo, Rossouw, & Margolis, 2013; Chinn et al., 2005); smoking cessation interventions that also limit weight gain are needed.

Since weight gain is a concern for older female smokers and given its relationship to smoking relapse, several research studies have explored the relationship of hormone therapy (HT) and weight gain among women smokers. Previous research has found that estrogen has been shown to have an appetitive-depressive effect in both animal and human studies (Alberti-Fidanza, Fruttini, & Maurizio, 1998; Blausteinb & Wade, 1976; Bowen & Grunberg, 1990) and could result in decreased weight gain. Given this finding, it is important to compare HT and non-HT women. Researchers have suggested that HT might reduce smoking-cessation increases in food intake and body weight in postmenopausal women.

To test this hypothesis, Allen, Brintnell, Hatsukami, & Reich (2004) investigated the effect of short-term (2-week) smoking abstinence on weight gain, energy intake, and physical activity in postmenopausal women currently taking and not taking HT. The participants were randomized to abstinence or continued smoking. The authors found that the abstainers gained more weight than the continued smokers; however, there were no significant differences between HT and non-HT abstainers, and between HT users and nonusers who continued smoking. Similar to this set of findings, in a study examining the effect of using nicotine replacement therapy (NRT) and hormone therapy (HT), during 2 weeks of smoking cessation on postmenopausal women's caloric intake and weight gain, the authors found no effects for HT and that women on NRT gained less weight than women on the placebo patch (Allen, Hatsukami, Brintnell, & Bade, 2005). Given that the above studies only had a two-week follow-up period, studies with longer follow-up periods are needed to examine the relationship between HT and post-cessation weight gain.

Furthermore, in postmenopausal women, the onset of hormonal changes tends to overlap with nicotine withdrawal symptoms (i.e., fluctuations in weight control). Given this overlap, it is posited that HT, which helps to counteract the negative effects of postmenopausal hormonal changes, could also help to alleviate some of the nicotine withdrawal symptoms during smoking cessation (Allen, Hatsukami, & Christianson, 2003). In a recent study, Jenks & Higgs (2011) examined the relationship between women dieters and non-dieters during smoking cessation and found that those who are dieting may experience greater withdrawal symptoms than those who are not dieting. Future research could examine the relationship between women dieters and non-dieters and HT users and nonusers during smoking cessation.

In summary, post-cessation weight gain appears to be a barrier for menopausal women, in addition to their menopause-related weight gain. Although there are significant health benefits to quitting smoking, even with some weight gain, weight gain could hinder some of the benefits of smoking cessation and interventions that address weight gain could help alleviate weight gain. However, given the few existing studies, there does not appear to be any evidence to suggest that HT significantly influences weight gain among women smokers. A lack of randomized trials limits the strength of these conclusions and additional research is necessary.

SUMMARY AND CONCLUSION

In summary, the existing literature shows no evidence to suggest that HT significantly influences smoking cessation outcomes, nicotine withdrawal symptoms, postcessation appetite change, or weight gain among women smokers. However, studies investigating the effects of HT on postcessation mood and affect have yielded mixed results. The findings regarding mood appear to be most significant in their relation to HT use in women smokers and the cessation and withdrawal process.

These findings should be interpreted in light of recent research identifying the monoamine oxidase inhibiting (MAOI) content of cigarette smoke and its well-established role in mood regulation, most notably its antidepressant effects. Specifically, the beta-carboline alkaloids in tobacco smoke have been found to act as inhibitors of human monoamine oxidase (e.g., Herraiz & Chaparro, 2005). Although the degree of MAOI activity in cigarette smoke has been found to vary with tobacco type (Lewis, Truman, Hosking, & Miller, 2012), the general antidepressant content in cigarette smoke has been found consistently, and animal research in this area may shed light on problematic processes in the clinical course of smoking, such as relapse (Guillem, Vouillac, Koob, Cador, & Stinus, 2008).

The MAOI content in cigarette smoke and its effects must also be viewed in light of hormonal changes that occur during the menopausal transition for women. The increased risk of major affective disorders ranging from mild cyclothymia to bipolar disorder, as well as unipolar major depressive disorder and dysthymia during perimenopause has been associated primarily with diminishing levels of estrogen. Although the specific neurobiological mechanism has yet to be identified for affective changes, it has been found that the volume of MAO-A is elevated in perimenopausal women, and that the increased MAO-A distribution coincides with declining estrogen levels (Rekkas, Wilson, Lee, Yogalingam, Sacher, Rusjan, Houle, Stewart, Kolla, Kish, Chiuccariello, Meyer, 2014). In fact, Rekkas and colleagues (2014) found that among women of perimenopausal age, MAO-A was elevated by 34% compared with women of reproductive age and by 16% compared with women in menopause.

To further highlight the critical role of estrogen and MAO levels in the brain, researchers have focused on the postpartum period—a time known to be associated with a sudden and significant drop in estrogen levels as well as with high risk for women to experience clinical depression (Sacher, Wilson, Houle, Rusjan, Hassan, Bloomfield, Stewart, & Meyer, 2010). Among postpartum women, Sacher and colleagues (2010) found that MAO-A volume measured in the prefrontal cortex, anterior cingulate cortex, anterior temporal cortex, thalamus, dorsal putamen, hippocampus, and midbrain was significantly elevated throughout the postpartum period.

Together such studies suggest that estrogen acts upon neurotransmitters in the same way as MAOI anti-depressants by increasing 5-HT-2 serotonin receptor binding as well as activating additional serotonin receptors and the overall concentration of serotonin. That is, it acts upon neurotransmitters in a manner similar to selective serotonin reuptake inhibitors (SSRI) anti-depressants. Estrogen likely inhibits re-uptake of serotonin by receptors and also inhibits the breakdown of the neurotransmitter thus increasing concentrations of it. It also increases norepinephrine binding to receptors and the turnover (breakdown and replacement) rate for norepinephrine. While estrogen may reduce MAO activity, progesterone may increase it. Progesterone also increases serum levels of norepinephrine, but it inhibits binding of it to receptors. Synthetic progestins appear to have a similar chemical action on neurotransmitters and have been directly linked to HT-associated depression (e.g., Sherwin, 1999).

Taken together, this information suggests a highly interesting and complicated role for HT in smoking cessation and whether or not HT users benefit from it in coping with the mood-related changes associated with cessation and withdrawal. HT's influence on mood may differ depending on the estrogen to progestin ratio included in it. This ratio content in turn, likely interacts with the phase in which women smokers are in the menopausal transition—perimenopause versus postmenopause—phases during which the levels of MAO have been shown to differ significantly.

Future Considerations

There are several important considerations in attempting to draw meaningful conclusions from these studies, as well as in designing studies for future research in this area; these include: (1) the use of pharmacological intervention, such as bupropion, varenicline, or nicotine replacement therapy as smoking cessation aids in clinical trials and how they interact with HT to influence withdrawal symptoms and cessation outcomes. Given the information regarding the MAO volume associated with the menopausal transition, perhaps such pharmacotherapies should be supplemented with antidepressants in older women smokers embarking on cessation. Future studies may determine if there is a beneficial role for MAOIs or SSRIs for these women. Indeed, the MAOI content in cigarette smoke may account for some of the findings regarding the lack of benefit women smokers often show with nicotine replacement therapies as compared to men. (2) the self-selection for HT use versus random assignment to receive HT among women participating in existing studies; and (3) the difficult distinction between mood, appetite/weight, and other physiological symptoms related to nicotine dependence or withdrawal, and those symptoms related to the menopausal transition.

References

Alberti-Fidanza, A., Fruttini, D., & Maurizio, S. (1998). Gustatory and food habit changes during the menstrual cycle. *International Journal for Vitamin and Nutrition Research, 68*, 149-153.

Allen, A. M., Kleppinger, A., Lando, H., & Oncken, C. (2013). Effect of nicotine patch on energy intake and weight gain in postmenopausal women during smoking cessation. *Eating Behaviors, 14*, 420-423.

Allen, S. S., Brintnell, D. M., Hatsukami, D., Reich, B. (2004). Energy intake and physical activity during short-term smoking cessation in postmenopausal women. *Addictive Behaviors, 29*, 947-951.

Allen, S.S., Hatsukami, D.K, Bade, T., & Center, B. (2004) Transdermal nicotine use in postmenopausal women: Does the treatment efficacy differ in women using and not using hormone replacement therapy? *Nicotine & Tobacco Research, 6*, 777-788.

Allen, S. S., Hatsukami, D. K., Brintnell, D. M., & Bade, T. (2005). Effect of nicotine replacement therapy on post-cessation weight gain and nutrient intake: A randomized controlled trial of postmenopausal female smokers. *Addictive Behaviors, 30*, 1273-1280.

Allen, S. S., Hatsukami, D. K., & Christianson, D. (2003). Nicotine withdrawal and depressive symptomatology during short-term smoking abstinence: A comparision of postmenopausal women using and not using hormone replacement therapy. *Nicotine & Tobacco Research, 5*, 49-59.

Barbaglia, G., Macia`, F., Comas, M., Sala, M., del Mar Vernet, M., Casamitjana, M., & Castells, X. (2009). Trends in hormone therapy use before and after publication of the Women's Health Initiative trial: 10 years of follow-up. *Menopause, 16*, 1061–1064.

Ben-Zaken Cohen, S., Pare, P., Man, S., & Sin, D. (2007). The growing burden of chronic obstructive pulmonary disease and lung cancer in women: examining sex differences in cigarette smoke metabolism. *American Journal of Respiratory and Critical Care Medicine, 176*, 113–120.

Blaustein, J. D., & Wade, G. N. (1976). Ovarian influences on meal patterns of female rats. *Physiology and Behavior*, 201-208.

Bowen, D. J., & Grunberg, N. E. (1990). Variations in food preference and consumption across the menstrual cycle. *Physiology and Behavior, 47*, 287-291.

Brandon, T. H. (1994). Negative affect as motivation to smoke. *Current Directions in Psychological Science, 3*, 33-37.

Caan, B., Coates, A., Schaefer, C., Finkler, L., Sternfeld, B., & Corbett, K. (1996). Women gain weight 1 year after smoking cessation while dietary intake temporarily increases. *Journal of the American Dietetic Association, 96*, 1150-1155.

Center for Disease Control; CDC, 2011

Chinn, S., Jarvis, D., Melotti, R., Luczynska, C., Ackermann-Liebrich, U., Anto, J. M., … Burney, P. (2005). Smoking cessation, lung function, and weight gain. A follow-up study. *The Lancet, 365*, 1629-1635.

Chlebowski, R., Anderson, G., Manson, J., Schwartz, A., Wakelee, H., Gass, M., . . . Kotchen, J. (2010). Lung Cancer Among Postmenopausal Women Treated With Estrogen Alone in the Women's Health Initiative Randomized Trial. *JNCI Journal of the National Cancer Institute, 102*, 1413–1421.

Copeland, A. L., Martin, P. D., Rash, C., Kendzor, D. E., & Geiselman, P. J. (2006). Predictors of attrition from smoking cessation treatment among pre- and postmenopausal weight-concerned women. *Eating Behaviors, 7*, 242-251.

Cropsey, K. L., McClure, L. A., Jackson, D. O., Villalobos, G. C., Weaver, M. F., & Stitzer. (2010). The impact of quitting smoking on weight among women prisoners participating in a smoking cessation intervention. *American Journal of Public Health, 100*, 1442-1448.

Duval, K., Prud'homme, D., Rabasa-Lhoret, R., Strychar, I., Brochu, M., Lavoie, J.M., & Doucet, É. (2013). Effects of the menopausal transition on dietary intake and appetite: A MONET Group Study. *European Journal of Clinical Nutrition, 68*, 271-6.

Filozof, C., Pinilla, M., & Fernandez-Cruz, A. (2004). Smoking cessation and weight gain. *Obesity Reviews, 5*, 95-103.

Gritz, E. R., Nielsen, I. R., & Brooks, L. A. (1996). Smoking cessation and gender: The influence of physiological, psychological, and behavioral factors. *Journal of the American Medical Women's Association, 51*, 35-42.

Guillem, K., Vouillac, C., Koob, G. F., Cador, M., & Stinus, L. (2008). Monoamine oxidase inhibition dramatically prolongs the duration of nicotine withdrawal-induced place aversion. *Journal of Biological Psychiatry, 63(2)*, 158-63.

Hale, G. E., Zhao, X., Hughes, C. L., Burger, H. G., Robertson, D. M., & Fraser, I. S. (2007). Endocrine features of menstrual cycles in middle and late reproductive age and the menopausal transition classified according to the Staging of Reproductive Aging Workshop (STRAW) Staging System. *Journal of Clinical Endocrinology & Metabolism: Clinical and Experimental, 92*, 3060–3067.

Hall, S. M., Ginsberg, D., & Jones, R. T. (1986). Smoking Cessation and Weight Gain. *Journal of Consulting and Clinical Psychology, 54*, 342-346.

Hatsukami, D., LaBounty, L., Hughes, J., & Laine, D. (1993). Effects of tobacco abstinence on food intake among cigarette smokers. *Health Psychology, 12*, 499-502.

Hatsukami, D., Skoog, K., Allen, S., & Bliss, R. (1995). Gender and the effects of different doses of nicotine gum on tobacco withdrawal symptoms. *Experimental and Clinical Psychopharmacology, 3*, 163-173.

Herraiz, T. & Chaparro, C. (2005). Human monoamine oxidase is inhibited by tobacco smoke: beta-carboline alkaloids act as potent and reversible inhibitors. *Biochemical and Biophysical Research Communications, 326(2)*, 378-86.

Jenks, R. A., & Higgs, S. (2011). Reactivity to smoking- and food-related cues in currently dieting and non-dieting young women smokers. *Journal of Psychopharmacology, 25*, 520-529.

Kasteridis, P. & Yen, S. T. (2012). Smoking cessation and body weight: evidence from the Behavioral Risk Surveillance Survey. *Health Services Research Journal, 47*, 1580-1602.

Kleppinger, A., Litt, M. D., Kenny, A. M., Oncken, C. A. (2010). Effects of smoking cessation on body composition in postmenopausal women. *Journal of Women's Health, 19*, 1651-1657.

Korhonen, T., Goodwin, A., Miesmaa, P., Dupuis, E. A., & Kinnunen, T. (2011). Smoking cessation program with exercise improves cardiovascular disease biomarkers in sedentary women. *Journal of Women's Health, 20*, 1051-1064.

Luo, J., Rossouw, J., Margolis, K. L. (2013). Smoking cessation, weight change, and coronary heart disease among postmenopausal women with and without diabetes. *Journal of the American Medical Association*, 310, 94-96.

Majumdar, Almasi, & Stafford, 2004 Majumdar, S., Almasi, E., & Stafford, R. (2004). Promotion and prescribing of hormone therapy after report of harm by the Women's Health Initiative. *Journal of the American Medical Association, 292*, 1983–1988.

Marshall, S., Clarke, C., Deapen, D., Henderson, K., Largent, J., Neuhausen, S., Stram, D. (2010). Recent breast cancer incidence trends according to hormone therapy use: The California Teachers Study cohort. *Breast Cancer Research, 12*, R4.

McVay, M. A. & Copeland, A. L. (2011). Smoking cessation in peri- and postmenopausal women: A review. *Experimental and Clinical Psychopharmacology, 19*, 192-202.

Michnovicz, J. J., Hershcopf, R. J., Naganuma, H., Bradlow, H. L., & Fishman, J. (1986). Increased 2-hydroxylation of estradiol as a possible mechanism for the anti-estrogenic effect of cigarette smoking. *The New England Journal of Medicine, 315*, 1305.

Nides, M., Rand, C., Dolce, J., Murray, R., O'Hara, P., Voelker, H., & Connett, J. (1994). Weight gain as a function of smoking cessation and 2-mg nicotine gum use among middle-aged smokers with mild lung impairment in the first 2 years of the lung health study. *Health Psychology, 13*, 354-361.

O'Hara, P., Connett, J. E., Lee, W. W., Nides, M., Murray, R., & Wise, R. (1998). Early and late weight gain following smoking cessation in the Lung Health Study. *American Journal of Epidemiology, 148*, 821.

Oncken, C., Cooney, J., Feinn, R., Lando, H., & Kranzler, H. R. (2007). Transdermal nicotine for smoking cessation in postmenopausal women. *Addictive Behaviors, 32*, 296-309.

Perkins, K. A. (2001). Smoking cessation in women: Special considerations. *CNS Drugs, 15*, 391-411.

Perkins, K. A. (1992). Effects of tobacco smoking on caloric intake. *British Journal of Addiction, 87*, 193-205.

Prior, J. C. (1998). Perimenopause: The complex endocrinology of the menopausal transition. *Endocrine Reviews, 19*, 397–428.

Rekkas, P. V., Wilson, A. A., Lee, V. W., Yogalingam, P., Sacher, J., Rusjan, P., Houle, S., Stewart, D. E., Kolla, N. J., Kish, S., Chiuccariello, L., Meyer, J. H. (2014). Greater monoamine oxidase a binding in perimenopausal age as measured with carbon 11-labeled harmine positron emission tomography. *JAMA Psychiatry, 71(8)*, 873-9.

Rossouw, J., Prentice, R., Manson, J., Wu, L., Barad, D., Barnabei, V., . . . Stefanick, M. (2007). Postmenopausal hormone therapy and risk of cardiovascular disease by age and years since menopause. *Jama, 297*, 1465–1477.

Sacher, J., Wilson, A. A., Houle, S., Rusjan, P., Hassan, S., Bloomfield, P. M., Stewart, D. E., Meyer, J. H. (2010). Elevated brain monoamine oxidase A binding in the early postpartum period. *Archives of General Psychiatry, 67(5)*, 468-74.

Sherwin, B. B. (1999). Progestogens used in menopause. Side effects, mood and quality of life. *Journal of Reproductive Medicine, 44(2 Suppl)*, 227-32.

Swan, G. E., Ward, M. M., Carmelli, D., & Jack, L. M. (1993). Differential rates of relapse in subgroups of male and female smokers. *Journal of Clinical Epidemiology, 46*, 1041-1053.

Tanko´, L. B., & Christiansen, C. (2004). An update on the antiestrogenic effect of smoking: A literature review with implications for researchers and practitioners. *Menopause, 11*, 104.

Wade, G. N. (1975). Some effects of ovarian hormones on food intake and body weight in female rats. *Journal of Comparative and Physiological Psychology, 88*, 183-193.

Wilson, P., Garrison, R., & Castelli, W. (1985). Postmenopausal estrogen use, cigarette smoking, and cardiovascular morbidity in women over 50. *New England Journal of Medicine, 313*, 1038–1043.

Wing, R. R., Matthews, K. A., Kuller, L. H., Neilahn, E. N., & Plantinga, P. L. (1991). Weight gain at the time of menopause. *Archives of Internal Medicine, 151*, 97-102.

Zang, E. A., & Wynder, E. L. (1996). Differences in lung cancer risk between men and women: Examination of the evidence. *Journal of the National Cancer Institute, 88*, 183.

In: Nicotine Dependence
Editor: Carlton L. Owens

ISBN: 978-1-63482-642-6

Chapter 2

A POSITIVE FEEDBACK LOOP OF SMOKING RISK

Heather A. Davis, Leila Guller, Elizabeth N. Riley and Gregory T. Smith
University of Kentucky, Lexington, KY, US

ABSTRACT

Recent research has suggested that certain personality traits and experiences predict and help maintain smoking behavior, thus leading to dependence and negative health consequences. A crucial challenge is to understand how people's early engagement in smoking behavior develops into problematic levels of use, despite their knowledge of the harms associated with smoking. In this chapter, we describe evidence indicating that there appears to be a positive feedback loop of increasing smoking and increasing risk for future smoking due to transactional processes involving high-risk personality traits, high-risk psychosocial learning, and early adolescent smoking behavior. *A striking aspect of the findings supporting this model is that there is no evidence of a corrective process, whereby early smoking experiences lead to reduced risk. The opposite appears to be the case: early smoking leads to greater personality-based and learning-based risk. Remarkably, this process occurs in a climate in which virtually everyone knows about the health risks of smoking.*

A Positive Feedback Loop of Smoking Risk

There is an extensive literature on the prevalence of smoking among adolescents, a prevalence that exists despite the widely publicized health risks. It may be useful to understand the process by which engagement in smoking occurs and is maintained over time among adolescents. Emerging research suggests that specific personality traits and experiences contribute to this process. After a brief review of the negative health effects of smoking, we will propose, based on empirical evidence, a model of risk for the adolescent onset of, and subsequent increases in, smoking. This model will involve a positive feedback loop, or vicious cycle, by which certain personality traits, combined with learned expectancies for reinforcement from smoking, predict and maintain actual smoking behavior.

Negative Health Consequences of Smoking

The negative health consequences of smoking are well documented. Cigarette smoking causes more than 480,000 deaths each year in the United States (CDC, 2010; U.S. Department of Health and Human Services, 2014; 2010). Smoking increases risk for cardiovascular disease, heart attacks, respiratory disease, and cancer in almost every area of the body (CDC, 2010). Smoking among adolescents is also prevalent, with rates increasing substantially across the adolescent years after age 12 (Chassin, Presson, Sherman, & Edwards, 1990). Early engagement in smoking increases the timeframe of exposure to the harmful effects of nicotine (Wills, Sandy, Yaeger, Cleary, & Shinar, 2001), the quantity of cigarettes smoked per day, and the likelihood of addiction to tobacco during adolescence and into adulthood (Chassin, Presson, Pitts, & Sherman, 2000). Finally, tobacco use has been linked to stunted physical growth in girls (Stice & Martinez, 2005).

Understanding of the process by which smoking behavior develops may inform intervention efforts and prevent these negative health effects from occurring. In the rest of this chapter, we present two complementary models of risk, the first emphasizing psychosocial learning and the second emphasizing personality. We then offer an integration of those two models. These models concern very early smoking behavior: our focus is on early adolescence. In each section, we present the model and then review empirical evidence supporting it.

PSYCHOSOCIAL LEARNING: POSITIVE FEEDBACK LOOP OF EXPECTANCIES

Our model of psychosocial learning emphasizes learning about possible positive effects of smoking. To the degree that individuals perceive there to be positive consequences to their smoking, smoking behavior is reinforced, thus leading to subsequent increases in the behavior.

Psychosocial learning about smoking can be assessed through formation of what are referred to as expectancies. Expectancies are thought to represent summaries of one's learning history about the outcomes of one's behavioral choices (Bolles, 1972; Tolman, 1932). Therefore, expectancies are understood to be proximal predictors of behavior: individuals engage in behaviors from which they expect reinforcement (Bolles, 1972; Tolman, 1932). Further, expectancies can provide evidence for memory-based associative learning (Goldman, Del Boca, & Darkes, 1999; Goldman, Reich, & Darkes, 2006).

From this perspective, smoking expectancies are understood to be learned associations indicating the expected consequences of smoking behavior. For example, high scores on a scale reflecting the expectancy that cigarette smoking facilitates positive social interactions are thought to reflect a strong learned association between smoking and social facilitation. Similarly, high scores on a scale reflecting the expectancy that smoking reduces negative feelings are thought to reflect a learned association between smoking and negative affect reduction.

Support for the psychosocial learning model of expectancies, and for the explicit measurement of smoking expectancies as markers of learned associations, has accrued over several decades (Bolles, 1972; Brandon & Baker, 1991; Lewis-Esquerre, Rodriguez, & Kahler, 2005; Myers, MacPherson, McCarthy, & Brown, 2003; Rotter, 1975; Tolman, 1932). Expectancies for reinforcement from smoking are thought to develop from a host of learning experiences that include modeling from others as well as direct experience. Because individuals pursue behaviors from which they expect reinforcement, expectancies for reinforcement from smoking at any one time should predict subsequent smoking, whether initiation of smoking in nonsmokers or increased smoking behavior in smokers. To the degree one has learned to expect reinforcement from smoking, one is more likely to smoke.

In the same vein, smoking behavior itself is an event that should influence subsequent expectancies. Each smoking event provides additional learning about the expected consequences of smoking. Thus, smoking behavior at any one time should predict subsequent expectancies. Although smoking has obvious and profound negative effects on health in the long run, to the degree that it does have immediate reinforcing properties, smoking behavior should predict subsequent increases in expectancies for reinforcement from smoking. If one perceives smoking to help reduce negative affect or to facilitate social experience, one's expectancies for those positive outcomes will be strengthened. Because immediate reinforcement has stronger influences on future behavior than delayed negative effects, our model holds that each smoking experience will be associated with a subsequent increase in expectancies for reinforcement from future smoking.

If both of these predictions, that smoking expectancies predict increases in smoking and smoking behavior predicts increases in smoking expectancies, are accurate, the implication is that there is a positive feedback loop taking place involving increasing smoking and increasing risk for future smoking over time. An alternative model would be that the negative effects of smoking would reduce positive expectancies for reinforcement from smoking. If that were true, engagement in smoking would provide a corrective experience that challenges any early held expectancies that smoking is reinforcing. Of these two models, we believe the positive feedback loop to be more accurate. The basis for this inference is that reinforcing effects of smoking are likely to be immediate, whereas punishing effects (such as health problems) are delayed, and thus less influential on subsequent behavior.

We recently tested this positive feedback loop hypothesis (Guller et al., in press-a). In a longitudinal study, we followed 1,906 children across three waves: spring of 5th grade, fall of 6th grade, and spring of 6th grade. Because middle school started in 6th grade for the study participants, we conducted our test during the transition to middle school. We studied children at a young age in order to better understand factors associated with the initiation of, and very early engagement in, smoking behavior.

We found that expectancies for reinforcement from smoking, held by elementary school children, predicted both onset of, and subsequent increases in, smoking behavior over the 18 month time frame (Guller et al., in press-a). It was striking that smoking expectancies predicted subsequent smoking onset: this finding suggests that the development of smoking expectancies at least partially occurs prior to smoking experience and thus is influenced by factors other than direct smoking experience. We also found that the prediction from smoking onset to subsequent smoking experience was positive (early smoking led to increased endorsement of expectancies for reinforcement from smoking). There was no evidence of a corrective process, in which early smoking led to less expectancy-based risk. Instead, these results demonstrate evidence supporting our model of a positive feedback loop in which smoking behavior increases expectancies for reinforcement from smoking, and smoking expectancies lead to more smoking behavior.

POSITIVE FEEDBACK LOOP BETWEEN SMOKING AND HIGH-RISK PERSONALITY

A large body of research supports the association between personality traits and cigarette smoking, particularly with respect to personality traits related to impulsive behavior (Bickel, Odum, & Madden, 1999; Bloom, Matsko, & Cimino, 2013; Doran, Cook, McChargue, & Spring, 2009; Mitchell, 2004; Perkins et al., 2008; Settles, Fischer, Cyders, Combs, Gunn, & Smith, 2012; Spillane, Smith, Kahler, 2010). One trait that appears to be particularly important for explaining smoking behavior is urgency, which is the tendency to act rashly when highly emotional (Cyders & Smith, 2008). Elevations in urgency predict smoking behavior (Billieux, Van der Linden, & Ceschi, 2007; Settles et al., 2012; Spillane et al., 2010). We next present a model describing urgency's role in the onset and maintenance of smoking behavior. To do so, we first consider basic aspects of normal personality change during early adolescence. We then review research describing urgency's role in predicting smoking behavior. After that, we present a model of personality change and introduce the hypothesis that smoking behavior can lead to changes in urgency. We then summarize our model of reciprocal prediction between urgency and smoking behavior.

Normative Personality Change in Early Adolescence

Personality change does appear to function as part of normal development in adolescents. During adolescence, two important aspects of personality, cognitive control and impulsivity, are thought to develop on different developmental trajectories. At least in part, this appears to be because the prefrontal cortex, which is important for control of impulses, develops more slowly than does the amygdala and ventral striatum, which are areas responsive to emotion and reward (Harden & Tucker-Drob, 2011). In fact, the socioemotional system (the amygdala and ventral striatum) is thought to develop early in adolescence, while the prefrontal cortex (cognitive control system) is not fully mature until early adulthood (Somerville, Jones, & Casey, 2010). This process is sometimes described as the dual systems model of adolescent development (Casey, Getz, & Galvan, 2008; Harden & Tucker-Drob, 2011; Steinberg, 2008, 2010; Somerville et al., 2010). Thus, early adolescence appears to be characterized by a well-developed responsiveness to emotion and reward, but a nascent capacity for cognitive control.

One obvious result of this developmental asynchrony is high levels of impulsive behavior observed in early adolescence (Steinberg, Albert, Cauffman, Banich, Graham, & Woolard, 2008). Overall, impulsivity declines linearly across adolescence and early adulthood and levels off once individuals reach their mid-twenties (Harden & Tucker-Drob, 2011; Steinberg et al., 2008). This is thought to occur because development of the cognitive control systems "catch up" to the socioemotional system.

Our model focuses on those adolescents who engage in smoking behavior early in adolescence. For reasons we describe below, we believe that early smoking interrupts the normal decline in impulsivity and leads to increases in the impulsivity-related trait of urgency (Riley, Davis, & Smith, in press).

Urgency Predicts Smoking Behavior

As described above, urgency refers to the disposition to engage in rash, impulsive behaviors when highly emotional (Cyders & Smith, 2008). There are two facets of this trait: negative urgency (the tendency to act rashly when distressed) and positive urgency (the tendency to act rashly when in an unusually positive mood: Cyders & Smith, 2008). Urgency predicts engagement in, and early onset of, several addictive behaviors including smoking, drinking, and binge eating among early adolescents (Guller et al, in press-b; Pearson, Combs, Zapolski, Smith, 2012; Settles et al., 2012; Settles, Zapolski, & Smith, 2014), and it does so above and beyond prediction from other impulsivity-related traits.

The inference that early adolescents are more likely to engage in these risky, maladaptive behaviors when highly emotional is consistent with the neurocognitive model of development described above (Harden & Tucker-Drobb, 2011) and with urgency theory (Cyders & Smith, 2008). Smoking and the other behaviors are thought to provide negative reinforcement in the form of distraction from distress and positive reinforcement in the form of social facilitation for smoking and drinking and pleasurable food consumption in the form of binge eating (Guller et al., in press-b; Heatherton & Baumeister, 1991; Hersh & Hussong, 2009; Small, Jones-Gotman, & Dagher, 2003).

Recent Research on Personality Change

To fully understand urgency's role in relation to smoking, it is important to consider an important, recent advance in the study of personality. Overall, personality is thought to be quite stable across the lifespan (Costa, Herbst, McCrae, & Seigler, 2000; Terracciano, Costa, & McCrae, 2006). Even the personality change in adolescence we described above is thought to reflect stability, in that the change is understood to be common to all adolescents, meaning that the rank order of individuals with respect to impulsivity remains stable in the midst of change (Harden & Tucker-Drobb, 2011). However, in recent years researchers have given more attention to the idea that, within that overall stability, personality can be dynamic and change over time (Bleidorn, 2012; Roberts, Walton, Viechtbauer, 2006; Roberts, Wood, & Caspi, 2008).

Studies of personality change find that going through periods of transition, such as the transition from adolescence to early adulthood (Bleidorn, 2012; Roberts et al., 2006; Roberts et al., 2008), can lead to personality change. As individuals transition into adulthood, new and different behaviors are required of them. For example, the need to be on time to work early in the morning each day, to pay bills on time, and to manage other adult responsibilities require an increase in conscientiousness. These new, more conscientious behaviors are reinforced, leading to greater investment in such behaviors and hence personality change in the direction of greater conscientiousness (Roberts et al., 2008).

However, people vary in how adaptively they manage this transition. Some adapt well to the change and experience positive change in personality, such as increases in conscientiousness (Roberts, Caspi, & Moffitt, 2001). Others respond to the transition challenge less well and engage in maladaptive behaviors, such as substance use, stealing from the workplace, and fighting with co-workers. They tend to experience negative changes in personality, such as increased negative emotionality and increased impulsiveness (Roberts et al., 2006). Thus, personality change, at least during periods of transition, does not necessarily occur in a common direction for all people.

An important question concerns specific mechanisms by which transitions lead to personality change. We believe a crucial factor is that the transitions require new behaviors, and the new behaviors change both one's typical experience and how one sees oneself (Riley et al., in press). When one engages in new, conscientious behaviors and they are reinforced, one's typical experience increasingly includes stable expressions of conscientiousness and one increasingly sees oneself as conscientious. When one engages in dysfunctional behaviors, such as substance use, and gets negative reinforcement from those behaviors (such as distraction from distress: Heatherton & Baumeister, 1991), one's typical experience increasingly includes dysfunctional acts and one increasingly sees oneself in that light.

Returning to risk for early smoking, our model holds that smoking behavior in early adolescence, which is an unusual, dysfunctional behavior, can lead to personality change in a maladaptive direction. Because smoking behavior can lead to reductions in negative affect or improved positive affect, we suggest it can lead to increases in the trait of urgency. Accordingly, we hypothesize a positive feedback loop between urgency and smoking in early adolescence: urgency predicts smoking onset and increased smoking, and we believe that smoking behavior leads to subsequent increases in urgency, thus heightening the personality-based risk for future smoking.

A Test of Reciprocal Prediction between Urgency and Smoking

Riley, Rukavina and Smith (2015) tested whether engagement in smoking, drinking, and binge eating predicted subsequent increases in urgency among adolescents over the course of a four year time frame, from 5th grade (the last year of elementary school) through 9th grade (the first year of high school). They found clear support for the hypothesis of reciprocal prediction. As has been shown before, urgency predicted increases in smoking, drinking, and binge eating throughout this four-year period. The study involved eight waves, and urgency at each wave predicted increases in those behaviors the following wave. In addition, they found that smoking, drinking, and binge eating at each wave predicted increases in urgency the following wave. Consistently, across the developmental transitions from elementary school to middle school and middle school to high school, they found reciprocal prediction between the personality trait of urgency and smoking behavior (as well as other addictive behaviors). These findings suggest a dangerous, reciprocal cycle of risk for youth high in the trait of urgency prior to adolescence.

Integration of the Positive Feedback Loop Processes

So far in this chapter, we have described two separate positive feedback loop risk processes, the first with respect to psychosocial learning and the second with respect to personality. Recent research suggests it may be possible to integrate these two processes into a single model. We consider this possibility next.

The Acquired Preparedness (AP) model provides one way to integrate personality and psychosocial learning processes to explain smoking risk (Davis, Riley, & Smith, in press; Smith & Anderson, 2001). Person-environment transaction theory (Caspi, 1993) holds that individuals with different personalities can perceive the same event very differently from each other. For example, a highly neurotic, high-strung person can perceive a social event as much more stressful than would an even-keeled, relaxed person. Smith and Anderson (2001) extended this idea to argue that, if people perceive a common event differently from each other, they are likely to learn different things from that common event. Smith, Williams, Cyders, and Kelley (2006) demonstrated this process in a laboratory longitudinal design.

It follows from the AP model that, as a function of differences in personality, individuals may be differentially prepared to acquire high-risk expectancies, such as expectancies for reinforcement from smoking. Numerous longitudinal studies have shown this process using the trait of urgency and different addictive behaviors. Consistently, urgency's prediction of subsequent addictive behavior involvement is mediated by urgency's influence on change in behavior-specific expectancies (Doran Khoddam, Sanders, Schweizer, Trim, & Myers, 2013; Pearson et al., 2012; Settles, Cyders, & Smith, 2010; Settles, Zapolski, & Smith, 2014). Combs and colleagues (2012) described an AP risk model for smoking in which they tested whether urgency would differentiate 5th grade smokers who had already smoked from those who had not, whether expectancies for reinforcement from smoking would differentiate the smokers from other 5th graders, and whether smoking expectancies would mediate the relationship between urgency and smoking behaviors. Their results indicated that the process by which urgency predicts smoking appears to be indirect and occurs through an individual's elevations in expectancies for reinforcement from smoking (Combs et al., 2012). Doran et al. (2013) showed the same process in a longitudinal design of late adolescents/young adults.

It is possible that our proposed positive feedback loop has multiple components. Perhaps high levels of urgency lead to increased expectancies for reinforcement from smoking, which then leads to increased smoking. In return, smoking directly leads to both increased expectancies (Guller et al., in press-b) and to increases in urgency in line with our personality change model (Riley et al., in press; Riley et al., 2015). Increases in urgency and in expectancies for reinforcement from smoking then further the risk for future smoking. This new, full, integrative model involving personality, expectancies, and smoking has not yet been tested.

Conclusion

Youth high in the risk factors of the personality trait of urgency and the learned expectancy that smoking is reinforcing are at risk to engage in a positive feedback loop of risk. Research shows that elevations in urgency predict the early onset of smoking and increased smoking in early adolescence; and, that early smoking, in turn, predicts subsequent increases in the trait of urgency. Similarly, elevations in the expectancy that smoking is reinforcing predicts early onset and early increases in smoking, and this early smoking behavior leads to yet stronger smoking expectancies. There is good reason to believe that the processes involving personality-based and expectancy-based risk can be integrated into an overall model of risk.

Strikingly, early engagement in smoking does not reduce future risk; it elevates it. Youth who smoke tend, subsequently, to associate smoking even more strongly with positive effects. In addition, their personalities tend to shift in a maladaptive direction. It is perhaps surprising that this process occurs in a society in which the health risks of smoking are widely known. Although virtually every adolescent presumably knows of the deleterious effects of smoking, the immediate reinforcement the behavior provides nevertheless leads to smoking behavior. Unfortunately, this early smoking appears to be part of a positive feedback loop of increasing risk to engage in this life-threatening behavior.

Stice, E., & Martinez, E. E. (2005). Cigarette smoking prospectively predicts retarded physical growth among female adolescents. *Journal of Adolescent Health, 37,* 363-370.

Terracciano, A., Costa, P. T., & McCrae, R. R. (2006). Personality plasticity after age 30. *Personality and Social Psychology Bulletin, 32*(8), 999-1009.

Tolman, E. C. (1932). *Purposive behavior in animals and men*. London England: Century/Random House UK.

U.S. Department of Health and Human Services (2014). The Health Consequences of Smoking—50 Years of Progress: A Report of the Surgeon General. Atlanta: U.S. Department of Health and Human Services, Centers for Disease Control and Prevention, National Center for Chronic Disease Prevention and Health Promotion, Office on Smoking and Health, 2014 [accessed 2014 Feb 6].

U.S. Department of Health and Human Services (2010). How Tobacco Smoke Causes Disease: What It Means to You. Atlanta: U.S. Department of Health and Human Services, Centers for Disease Control and Prevention, National Center for Chronic Disease Prevention and Health Promotion, Office on Smoking and Health, 2010 [accessed 2014 Feb 6].

Wills, T. A., Sandy, J. M., Yaeger, A. M., Cleary, S. D., & Shinar, O. (2001). Coping dimensions, life stress, and adolescent substance use: A latent growth analysis. *Journal of Abnormal Psychology, 110,* 309-323.

In: Nicotine Dependence
Editor: Carlton L. Owens

ISBN: 978-1-63482-642-6

Chapter 3

IMPACT OF ACTIVE AND PASSIVE WATER PIPE TOBACCO SMOKING

Hans Schweisfurth*
Institute for Pulmonary Research (IPR), Cottbus, Germany

ABSTRACT

Approximately 100 million people smoke water pipe (shisha, hookah, narghile) every day. This trend is increasing in the Middle East, US and in Europe especially in adolescents. The reason for the widely distributed use of water pipe smoking is nurtured by the misleading assumption that this form of smoking is less harmful than cigarette smoking. Studies revealed that tobacco smoke from water pipe, depending on the used tobacco and charcoal, contains higher concentrations of nicotine, carbon monoxide, carboxy hemoglobin, heavy metals, carcinogens and fine dust particles compared to cigarette smoke. Similar to cigarette smoking, water pipe tobacco smoking causes a severe health risk by second hand smoke, because the side-stream smoke of the water pipe contains higher concentrations of polycyclic aromatic hydrocarbons, aldehydes and carbon monoxide compared to the side-stream smoke of cigarettes. According to studies, active and passive water pipe tobacco smoking was significantly associated with lung

* Correspondence: Hans Schweisfurth, MD, Professor of Medicine, Institute for Pulmonary Research (IPR), Walther-Rathenau-Strasse 11, 03044 Cottbus, Germany, Email: ipr@pulmologisches-forschungsinstitut.de

cancer, chronic obstructive lung diseases, asthma, cardiovascular effects, periodontal diseases and low birth-weight.

Keywords: water pipe tobacco smoking, second-hand smoking, hooka, shisha, prevalence, toxicity, cancer, nicotine, cotinine, carboxy hemoglobin, respiratory diseases

INTRODUCTION

Each year, almost 6 million people die prematurely from the effects of tobacco smoking worldwide [1]. Low- and middle-income countries are the most severely affected. Tobacco caused deaths are projected to decline by 9% between 2002 and 2030 in high-income countries, but will double from 3.4 million to 6.8 million in low- and middle-income countries [2, 3]. In the last years the scientific interest is increasing in the effects of water pipe smoking [4]. Approximately 100 million people smoke water pipe daily, and it is a growing trend among young people in particular. Healthcare professionals worldwide need to be aware of this potential threat [5 - 15].

A water pipe consists of head, body, water bowl, and hose (Figure 1). Holes in the bottom of the head allow smoke to pass into the body's vertical tube. This tube is dipped into the water of the bowl. The hose exits from the top of the water bowl and ends with a mouthpiece. The tobacco is placed into the head and does not burn in a self-sustaining manner. The charcoal is placed atop the tobacco-filled head. When the charcoal is lit, a smoker inhales through the hose, creating a vacuum above the water, and drawing air through the body and over the tobacco and charcoal. Having passed over the charcoal, the heated air, which now also contains charcoal combustion products, passes through the tobacco, and the mainstream smoke aerosol is produced. The smoke passes through the water pipe body, bubbles through the water in the bowl, and is inhaled by the smoker who needs a high suction pressure and deep inspiratory effort to overcome the resistance of the whole system.

Most likely the water pipe has evolved from a coconut in India. There it is called "Narjil" just like the coconut. Hemp leafs were burnt in it and the smoke was inhaled through a straw [16-18].

In 1492 Christopher Columbus reported on tobacco smoking on the American continent [19]. A century later it was described that the indigenous population was smoking leaves through a small Y-shaped wooden pipe, named Tobago [20].

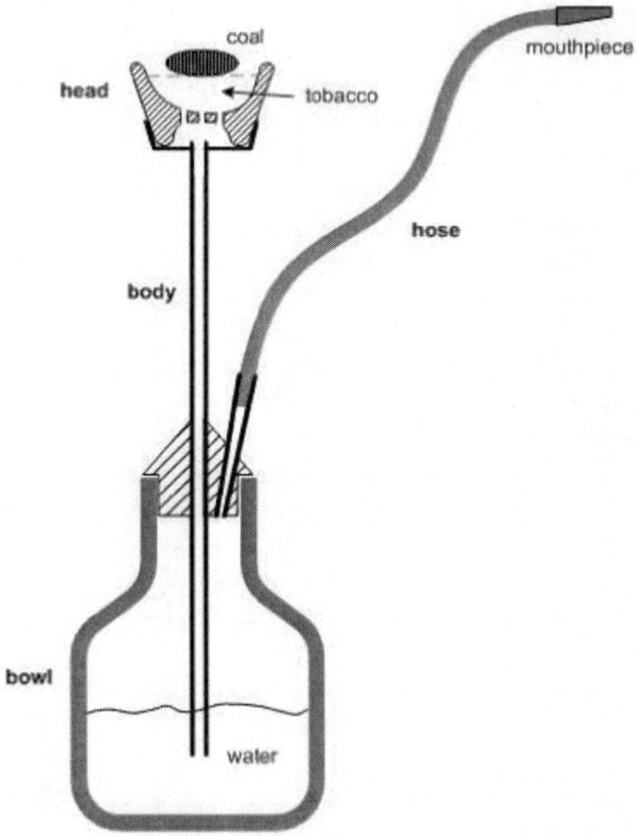

Figure 1. Water pipe cross-section view. The smoker draws the smoke from the tobacco and charcoal burning head through the water by the long hose [16, 106].

Between the 16th and 17th century mainly glass was used to manufacture water pipes in Turkey [17]. By 1700, the Turks introduced the water pipe through Egypt to the Middle East and Africa [18]. The Ottomans added a clay pipe and a mouthpiece to the water pipe [17]. This type of water pipe is now used in the Mediterranean, North Africa, parts of Asia as well as in the Western World [18].

The water pipe is called “Shisha”, “Boory” or “Goza” in Egypt and Saudi Arabia, “Narghile”, “Nargile” or “Arghile” in Israel, Jordan, Lebanon, Syria, “Hooka” in Africa and “Hubble Bubble” in the Indian subcontinent [7]. There are three different types of water pipes, which differ mainly in size and type of the water tank. The “Goza” is the oldest form of a metal water tank with a volume of 200-500 ml. The “Shisha” has a water tank which is made of glass, ceramics, rock crystal, brass or silver and has a capacity of about 1,000 ml [18]. The tobacco is deposited in a clay or metal bowl and covered with a metal screen or perforated aluminum foil. Then burning charcoal is placed on it. The smoke is inhaled by a column of smoke, which is usually equipped with a pressure valve [21].

"Maassel" is most commonly used as tobacco, which is enriched with fruit or mint [18]. It differs from cigarette tobacco by a higher proportion of humectant [21]. Other types of tobacco contain less sweeteners and are called "Ajamy" "Tutun "or" Jurak" [22]. But also marijuana, herbal shisha and hashish is used by young adults [23].

The toxicity of all used forms of tobacco is different [22]. In contrast to warnings on cigarette packages, adequate information on water pipe tobacco is often not available [24].

Besides tobacco, also some flavored non-tobacco products are used for water pipe smoking. Experimentally it was shown, that these tobacco-free products have the same quantities of toxicants, but only the nicotine level was very low. These results contradict advertising messages that the tobacco-free preparations are a healthy alternative to tobacco products [25].

PREVALENCE

The water pipe is used in India, Pakistan, China, in the eastern Mediterranean, Arab countries, Europe and North America (Table 1) [26 - 32]. A survey in Kuwait revealed that 57% of men and 69% of women had smoked water pipe [33]. In a recent study from Kuwait, the prevalence was reported to be 45% [34]. According to a comprehensive study (n = 196,268) the entire rate of cigarette and water pipe smokers amounts to 16.9% in Saudi Arabia [35]. Studies revealed, that the water pipe is used by 4 to 15% of smokers in Saudi Arabia. In Oman, 26.6% of smokers also smoke water pipe [36]. In Syria women tend more to water pipe smoking than to cigarette smoking [37].

Water pipe smoking is widespread mainly among youth [38]. In the Middle East the prevalence is estimated to 6-34% [8]. In most Gulf countries the rate of water pipe smokers among the 13 to 15 year old teenagers was higher than that of cigarette smokers [39]. In Egypt 26% of boys and 5% of girls smoke water pipe [40]. In Israel 22% of adolescents aged 12 to 18 years smoke water pipe at least once per week [41]. In Iran, the starting age was 16 years on average [42]. In Iraq 8% of the adolescents who smoked water pipe initiated water pipe smoking at the age under 10 years [43].

Table 1. Summary of epidemiological studies regarding prevalence of water pipe smoking [16, 65]. m: male, f: female, y: years

Region	Population	N	Prevalence of water pipe smoking	Reference
Tunisia 2003	School children, 52% m, age 13-17 y	1,569	19% ever 5% current	Harrabi 2010 [135]
Egypt 2002-2003	School students, 65% m, start with 12 y	635	19% (26% m, 5% f)	Gadalla 2003 [40]
Lebanon 2003-2005	Total population	37,579	4% (57% f) mean age 36 y	Riachy 2008 [136]
Lebanon 2002-2003	School students, 45% m, mean age15 y	2,443	26% [1] 30% [2]	Tamim[1] 2007 [137], El-Roueiheb [2] 2008 [45]
Lebanon	University students, 52% m, start with 16 y	416	28% current	Chaaya 2004 [138]
Lebanon 2003	Pregnant, mean age 27 y	864	before pregnancy 7%, during pregnancy 4%	Chaaya 2004 [67]
Lebanon 2001-2002	School students, 57% m, mean age 16 y, start with 5 y	1,461	24% >1/week 14% occasionally	Zoughaib 2004 [139]
Lebanon 2000	Pregnant women, mean age 28 y	576	18% ever during pregnancy 6%	Chaaya 2003 [66]
Lebanon 2000-2001	University students, 41% m, mean age 21 y	1,964	21%	Tamim 2003 [140]
Lebanon 1997	Adults, mean age 40 y	727	9% current	Baddoura 2001 [141]
Syria 2006-2007	Medical students, mean age 21 y	570	24% (30% m, 13% f)	Almerie 2008 [47]
Syria 2004	Adults, 45% m, mean age 35 y	2,038	1% daily 11% occasionally (20% m, 5% f)	Ward 2006 [142]

Table 1. (Continued)

Region	Population	N	Prevalence of water pipe smoking	Reference
Syria 2004	Adults, 46% m, mean age 34 y	1,021	9% (16% m, 4% f)	Maziak 2005 [143]
Syria 2003	University students, 47% m, mean age 22 y	587	15 % (26% m, 5% f) occasionally 7% daily 20% (30% m, 12% f)[1]	Maziak 2004 [31, 144, 145] Afsar[1] 2005 [37]
Turkey	University students	645	33% (42% m, 20% f)	Porzrazoglu 2010 [51]
Iran	University students Start with a mean age of 16 y	1,024	43% current 19% last 30 days	Sabahy 2011 [42]
Iraq 2012	School students age 13-18 y	1,750	4,8%	Hussain 2013 [43]
Middle East	Hospitalized patients with acute coronary syndrome	6,704	4%	Al Suwaidi 2012 [105]
Gulf Cooperation Council 2001-2004 (Kuwait, Bahrain, Saudi-Arabia, Qatar, UAE, Oman)	Students, age 13-15 y	32,356	Bahrain 15% Kuwait 16% Oman 9% Qatar 14% Saudi Arabia 10% UAE 15% Yemen 15%	Al-Mulla 2008 [39]
Bahrain	Secondary school-boys, mean age 17 y	600	13%	Al-Haddad 2003 [146]
Kuwait, Bahrain 2000-2001	Physicians, 65% m, mean age 45 y	1,565	Kuwait 12% (17% m, 3% f) Bahrain 6% (9% m, 3% f)	Behbehani 2004 [147]
Kuwait 1996	Governmental officials	3,859	63% (57% m, 69% f) ever	Memon 2000 [33]
UAE 2005	University students, 39% m, mean age 21 y	1,057	6% (26% f)	Mandil 2007 [148]
Oman 2003	School students, 51% m,	1,962	29% ever 10% (16% m, 3% w) current	Al-Lawati 2008 [36]

[16] Schweisfurth H, Kanaan A. Review article: Prevalence and health effects of water pipe (Shisha) tobacco smoking. *Glob. J. Pathol.* Microbiol. 2013, 1: 75-84.

[17] Gerçek G. Narghiles. Antika. The Turkish Journal of Collectable Art, February 1986, 11.

[18] World Health Organization: Regional Office for the Eastern Mediterranean Tobacco use in shisha: studies on waterpipe smoking in Egypt / WHO Regional Office for the Eastern Mediterranean, 2006.

[19] Benzoni. Travels in the Americas (1542–1556). 1565. Quoted from 1911 Encyclopedia, love to know article on tobacco. Available online at: http: //20.1911encyclopedia.org/T/TO/TOBACCO.htm

[20] Oviedo. Historia de las Indias Occidentales, Salamanca, 1535. Quoted from 1911. Encyclopedia. Love to know article on tobacco. Available online at: http://20.1911encyclopedia.org/T/TO/TOBACCO.htm

[21] Deutsches Krebsforschungszentrum (Ed.): Wasserpfeife – die süße Versuchung. Heidelberg, Germany, 2008.

[22] Chaouachi K. Assessment of narghile (shisha, hookah) smokers actual exposure to toxic chemicals requires further sound studies. *Libyan J.* Med 2011; 6: 5934.

[23] Sutfin EL, Song EY, Reboussin BA, Wolfson M. What are young adults smoking in their hookahs? A latent class analysis of substances smoked Addict Behav 2014 Jul; 39 (7): 1191-6. doi: 10.1016/ j.addbeh.2014.03.020. Epub 2014 Mar 26.

[24] Nakkash R, Khalil J. Health warning labelling practices on narghile (shisha, hookah) waterpipe tobacco products and related accessories. *Tob Control* 2010; 19: 235-239.

[25] Shihadeh A, Salman R, Jaroudi E et al. Does switching to a tobacco-free waterpipe product reduce toxicant intake? A crossover study comparing CO, NO, PAH, volatile aldehydes, "tar" and nicotine yields. *Food Chem. Toxicol.* 2012 May; 50 (5): 1494-1498. Doi: 10.1016/j.fct.2012.02.041. Epub 2012 Mar 1.

[26] El-Hakim IE, Uthman MA. Squamous cell carcinoma and keratoacanthoma of the lower lip associated with "Goza" and "Shisha" smoking. *Int. J. Dermatol.* 1999; 38: 108–110.

[27] Israel E, El-Setouhy M, Gadalla S et al. Water pipe (Sisha) smoking in cafes in Egypt. *J Egypt Soc Parasitol* 2003; 33: 1073–1085.

[28] Lubin JH, Qiao YL, Taylor PR et al. Quantitative evaluation of the radon and lung cancer association in a case control study of Chinese tin miners. *Cancer Res.* 1990; 50: 174–180.

[29] Lubin JH, Li JY, Xuan XZ et al. Risk of lung cancer among cigarette and pipe smokers in southern China. *Int. J. Cancer* 1992; 51: 390–395.

[30] Maziak W, Eissenberg T, Klesges RC et al. Adapting smoking cessation interventions for developing countries: a model for the Middle East. Int. *J. Tuberc. Lung. Dis*. 2004; 8: 403–413.

[31] Maziak W, Fouad MF, Hammal F, Asfar T. Prevalence and characteristics of narghile smoking among university students in Syria. *Int. J. Tuberc. Lung. Dis.* 2004; 8: 882–889.

[32] Maziak W, Rastam S, Eissenberg T et al. Gender and smoking status-based analysis of views regarding waterpipe and cigarette smoking in Aleppo, *Syria. Prev. Med*. 2004; 38: 479–484.

[33] Memon A, Moody PM, Sugathan TN et al. Epidemiology of smoking among Kuwaiti adults: prevalence, characteristics, and attitudes. Bull *World Health Organ* 2000; 78: 1306-1315.

[34] Mohammed HR, Zhang Y, Newman IM, Shell DF. Waterpipe smoking in Kuwait. East Mediterr Health J. 2010; 16: 1115-1120.

[35] Al-Turki KA, Al-Baghli NA, Al-Ghamdi AJ et al. Prevalence of current smoking in Eastern province, Saudi Arabia. *East Mediter Health J.* 2010; 16: 671-676.

[36] Al-Lawati JA, Muula AS, Hilmi SA, Rudatsikira E. Prevalence and determinants of waterpipe tobacco use among adolescents in Oman. *Sultan Qaboos Univ Med J*. 2008; 8: 37-43.

[37] Asfar T, Ward KD, Eissenberg T, Maziak W. Comparison of patterns of use, beliefs, and attitudes related to waterpipe between beginning and established smokers. *BMC Public Health* 2005; 25, 5: 19.

[38] Maziak W, Nakkash R, Bahelah R et al. Tobacco in the Arab world: old and new epidemics amidst policy paralysis. *Health Policy Plan* 2014 Sep; 29(6): 784-794. Epub 2013 Aug 19.

[39] Al-Mulla AM, Helmy SA, Al-Lawati J et al. Prevalence of tobacco use among students aged 13-15 years in Health Ministers' Council/Gulf Cooperation Council Member States, 2001-2004. *J. Sch. Health* 2008, 78: 337-343.

[40] Gadalla S, Aboul-Fotouh A, El-Setouhy M et al. Prevalence of smoking among rural secondary school students in Qualyobia governorate. *J. Egypt. Soc. Parasitol* 2003; 33: 1031-1050.

[41] Varsano S, Ganz I, Eldor N, Garenkin M. Water-pipe tobacco smoking among school children in Israel: frequencies, habits, and attitudes. *Harefuah* 2003; 142: 736–741, 807.

[42] Sabahy AR, Divsalar K, Bahreinifar S et al. Waterpipe tobacco use among Iranian university students: correlates and perceived reasons for use. *Int. J. Tuberc Lung. Dis.* 2011; 15: 844-847.

[43] Hussain HY, Abdulsatar BA. Prevalence and determinants of tobacco use among Iraqi adolescents: Iraq GYTS 2012. Tob Induc Dis 2013 Jun 28; 11(1):14. doi: 10.1186/1617-9625-11-14. eCollection 2013.

[44] Amin TT, Amr MA, Zaza BO, Kaliyadan F. Predictors of waterpipe smoking among secondary school adolescents in Al Hassa, Saudi Arabia. *Int. J. Behav. Med.* 2012 Sep; 19(3): 324-335. doi: 10.1007/s12529-011-9169-2.

[45] El-Roueiheb Z, Tamim H, Kanj M et al. Cigarette and waterpipe smoking among Lebanese adolescents, a cross-sectional study, 2003-2004. Nicotine Tob. Res. 2008; 10: 309-314.

[46] Saade G, Warren CW, Jones NR, Mokdad A. Tobacco use and cessation counseling among health professional students: Lebanon Global Health Professions Student Survey. *J. Med. Liban* 2009; 57: 243-247.

[47] Almerie MQ, Matar HE, Salam M et al. Cigarettes and waterpipe smoking among medical students in Syria: a cross-sectional study. *Int. J. Tuberc. Lung. Dis.* 2008; 12: 1085-1091.

[48] Jensen PD, Cortes R, Engholm G et al. Waterpipe use predicts progression to regular cigarette smoking among Danish youth. Subst. *Use Misuse* 2010; 45: 1245-1261.

[49] Maziak W. The waterpipe: time for action. *Addiction* 2008; 103: 1763-1767.

[50] Chéron-Launay M, Baha M, Mautrait C et al. Identifying addictive behaviors among adolescents: a school-based survey. *Arch. Pediatr.* 2011; 18: 737-744.

[51] Poyrazoğlu S, Sarli S, Gencer Z, Günay O. Waterpipe (narghile) smoking among medical and non-medical university students in Turkey. *Ups J Med Sci* 2010; 115: 210-216.

[52] Mach P, Zielińska-Danch W, Goniewicz MŁ et al. Waterpipe smoking among adolescents in Silesia region of Poland. *Przegl Lek* 2010; 67: 1021-1024.

[53] Pärna K, Usin J, Ringmets I. Cigarette and waterpipe smoking among adolescents in Estonia: HBSC survey results, 1994-2006. *BMC Public* Health 2008; 8: 392.

[54] Dugas E, Tremblay M, Low NC et al. Water-pipe smoking among North American youth. *Pediatrics* 2010; 125: 1184-1189.

[55] Primack BA, Walsh M, Bryce C, Eissenberg T. Water-pipe tobacco smoking among middle and high school students in Arizona. *Pediatrics* 2009, 123: e282-288.

[56] Cobb CO, Ward KD, Maziak W et al. Waterpipe tobacco smoking: an emerging health crisis in the United States. *Am. J. Health Behav*. 2010; 34: 275-285.

[57] Eissenberg T, Ward KD, Smith-Simone S, Maziak W. Waterpipe tobacco smoking on a U.S. College campus: prevalence and correlates. *J. Adolesc Health* 2008; 42: 526-529.

[58] Primack BA, Sidani J, Agarwal AA et al. Prevalence of and associations with waterpipe tobacco smoking among US university students. *Ann. Behav. Med.* 2008; 36: 81-86.

[59] Sutfin EL, McCoy TP, Reboussin BA et al. Prevalence and correlates of waterpipe tobacco smoking by college students in North Carolina. *Drug Alcohol Depend* 2011; 115: 131-136.

[60] Jamil H, Janisse J, Elsouhag D et al. Do household smoking behaviors constitute a risk factor for hookah use? *Nicotine Tob.* Res 2011; 13: 384-388.

[61] Martinasek MP, McDermott RJ, Martini L. Waterpipe (hookah) tobacco smoking among youth. *Curr. Probl. Pediatr. Adolesc. Health Care* 2011; 41: 34-57.

[62] Ward KD, Vander Weg MW, Relyea G et al. Waterpipe smoking among American military recruits. *Prev. Med.* 2006; 43: 92-97.

[63] Primack BA, Fertman CI, Rice KR et al. Waterpipe and cigarette smoking among college athletes in the United States. *J. Adolesc Health* 2010; 46: 45-51.

[64] Jackson D, Aveyard P. Waterpipe smoking in students: prevalence, risk factors, symptoms of addiction, and smoke intake. Evidence from one British university. *BMC Public Health* 2008; 8: 174.

[65] Akl EA, Gunukula SK, Aleem S et al. The prevalence of waterpipe tobacco smoking among the general and specific populations: a systematic review. *BMC Public Health* 2011; 11: 244.

[66] Chaaya M, Awwad J, Campbell OM et al. Demographic and psychosocial profile of smoking among pregnant women in Lebanon: public health implications. *Matern Child Health J.* 2003; 7: 179-186.

[67] Chaaya M, Jabbour S, El-Roueiheb Z et al. Knowledge, attitudes, and practices of argileh (water pipe or hubble-bubble) and cigarette smoking among pregnant women in Lebanon. *Addict Behav* 2004; 29: 1821-1831.

[68] Chaouachi K. Hookah (shisha, narghile) smoking and environmental tobacco smoke (ETS). A critical review of the relevant literature and the public health consequences. Int. J. Environ. Res. Public Health 2009; 6: 798–843.

[69] Jawaid A, Zafar AM, Rehman TU et al. Knowledge, attitudes and practice of university students regarding waterpipe smoking in Pakistan. *Int J Tuberc Lung Dis* 2008; 12: 1077-1084.

[70] Maziak W, Ward KD, Eissenberg T. Interventions for waterpipe smoking cessation. *Cochrane Database Syst. Rev.* 2007 Oct 17; 4: CD005549.

[71] Roskin J, Aveyard P. Canadian and English students' beliefs about waterpipe smoking: a qualitative study. BMC Public Health 2009; 9: 10.

[72] Shafagoj YA, Mohammed FI, Hadidi KA. Hubble-bubble (water pipe) smoking: levels of nicotine and cotinine in plasma, saliva and urine. *Int. J. Clin. Pharmacol. Ther.* 2002; 40: 249–255.

[73] Singh PN, Neergaard J, Job JS et al. Differences in health and religious beliefs about tobacco use among waterpipe users in the rural male population of Egypt. *J. Relig. Health* 2012 Dec; 51(4): 1216-1225. doi: 10.1007/s10943-010-9431-y.

[74] Smith SY, Curbow B, Stillman FA. Harm perception of nicotine products in college freshmen. *Nicotine Tob. Res.* 2007; 9: 977-982.

[75] Smith-Simone S, Maziak W, Ward KD, Eissenberg T. Waterpipe tobacco smoking: knowledge, attitudes, beliefs, and behavior in two U.S. samples. Nicotine Tob Res 2008; 10: 393-398.

[76] Al-Naggar RA, Bobryshev YV. Shisha smoking and associated factors among medical students in Malaysia. *Asian Pac J. Cancer Prev.* 2012; 13(11): 5627-5632.

[77] Nakkash R, Khalil J, Afifi R. The rise in narghile (shisha, hookah) waterpipe tobacco smoking: A qualitative study of perceptions of smokers and non smookers. *BMC Public Health* 2011; 11: 315.

[78] Al Mutairi SS, Shihab-Eldeen AA, Mojiminiyi OA, Anwar AA. Comparative analysis of the effects of hubble-bubble (sheesha) and cigarette smoking on respiratory and metabolic parameters in hubble-bubble and cigarette smokers. *Respirology* 2006; 11: 449–455.

[79] Al Rashidi M, Shihadeh A, Saliba NA. Volatile aldehydes in the mainstream smoke of the narghile waterpipe. *Food Chem Toxicol* 2008; 46: 3546-3549.

[80] Schubert J, Kappenstein O, Luch A, Schulz TG. Analysis of primary aromatic amines in the mainstream waterpipe smoke using liquid

chromatography-electrospray ionization tandem mass spectrometry. *J. Chromatogr A* 2011; 1218: 5628-5637.

[81] Sepetdjian E, Shihadeh A, Saliba NA. Measurement of 16 polycyclic aromatic hydrocarbons in narghile waterpipe tobacco smoke. *Food Chem Toxicol* 2008; 46: 1582-1590.

[82] Sepetdjian E, Saliba N, Shihadeh A. Carcinogenic PAH in waterpipe charcoal products. *Food Chem. Toxicol.* 2010; 48: 3242-3245.

[83] Shihadeh A. Investigation of mainstream smoke aerosol of the argileh water pipe. *Food Chem. Toxicol.* 2003; 41: 143-152.

[84] Shihadeh A, Saleh R: Polycyclic aromatic hydrocarbons, carbon monoxide, "tar", and nicotine in the mainstream smoke aerosol of the narghile waterpipe. *Food Chem. Toxicol.* 2005; 43: 655-661.

[85] Shihadeh AL, Eissenberg TE. Significance of smoking machine toxicant yields to blood-level exposure in waterpipe tobacco smokers. Cancer Epidemiol *Biomarkers Prev* 2011 Nov; 20(11): 2457-2460. doi: 10.1158/1055-9965.EPI-11-0586. Epub 2011 Sep 13.

[86] Al Ali R, Rastam S, Ibrahim I et al. A comparative study of systemic carcinogen exposure in waterpipe smokers, cigarette smokers and non-smokers. Tob Control 2013 Sep 2. doi: 10.1136/tobaccocontrol-2013-051206. [Epub ahead of print].

[87] Monn Ch, Kindler P, Meile A, Brändli O. Ultrafine particle emissions from waterpipes. *Tob. Control.* 2007; 16: 390-393.

[88] Schubert J, Hahn J, Dettbarn G et al. Mainstream smoke of the waterpipe: Does this environmental matrix reveal as significant source of toxic compounds? *Toxicol. Lett.* 2011; 205: 279-284.

[89] Khabour OF, Alsatari ES, Azab M et al. Assessment of genotoxicity of waterpipe and cigarette smoking in lymphocytes using the sister-chromatid exchange assay: a comparative study. *Environ. Mol. Mutagen* 2011; 52: 224-228.

[90] El-Setouhy M, Loffredo CA, Radwan G et al. Genotoxic effects of waterpipe smoking on the buccal mucosa cells. *Mutat. Res.* 2008; 655: 36-40.

[91] Warnakulasuriya S. Waterpipe smoking, oral cancer and other oral health effects. *Evid. Based Dent.* 2011; 12: 44-45.

[92] Maziak W. The waterpipe: an emerging global risk for cancer. Cancer Epidemiol 2013 Feb; 37(1): 1-4. doi: 10.1016/j.canep.2012.10.013. Epub 2012 Nov 26.

[93] Akl EA, Gaddam S, Gunukula SK et al. The effect of waterpipe tobacco smoking on health outcomes: a systematic review. *Int. J. Epidemiol* 2010; 39: 834-857.

[94] Onder M, Özta, Arnavut O. Nargile (hubble-bubble) smoking–induced hand eczema. *Int. J. Dermatol* 2002; 41: 771-772.

[95] Hakim F, Hellou E, Goldbart A et al. The acute effects of water-pipe smoking on the cardiorespiratory system. *Chest*. 2011; 139: 775-781.

[96] Van der Vaart H, Postma DS, Timens W et al. Acute effects of cigarette smoking on inflammation in healthy intermittent smokers. *Respir. Res.* 2005; 6: 22

[97] Al-Kubati M, Al-Kubati AS, al'Absi M, Fiser B. The short-term effect of water-pipe smoking on the baroreflex control of heart rate in normotensives. *Auton. Neurosci*. 2006; 126-127: 146–149.

[98] Blank MD, Cobb CO, Kilgalen B et al. Acute effects of waterpipe tobacco smoking: a double-blind, placebo-control study. *Drug. Alcohol.* Depend 2011; 116: 102-109.

[99] Eissenberg T, Shihadeh A. Waterpipe tobacco and cigarette smoking: direct comparison of toxicant exposure. *Am. J. Prev. Med.* 2009; 37: 518–523.

[100] Shafagoj YA, Mohammed FI. Levels of maximum end-expiratory carbon monoxide and certain cardiovascular parameters following hubble-bubble smoking. *Saudi Med J.* 2002; 23: 953–958.

[101] Shaikh RB, Vijayaraghavan N, Sulaiman AS et al. The acute effects of waterpipe smoking on the cardiovascular and respiratory systems. *J. Prev. Med. Hyg* 2008; 49: 101–107.

[102] Cobb CO, Sahmarani K, Eissenberg T, Shihadeh A. Acute toxicant exposure and cardiac autonomic dysfunction from smoking a single narghile waterpipe with tobacco and with a "healthy" tobacco-free alternative. *Toxicol Lett* 2012 Nov 23; 215(1): 70-75. Doi: 10.1016/j.toxlet.2012.09.026. Epub 2012 Oct 8.

[103] Kadhum M, Jaffery A, Haq A et al. Measuring the acute cardiovascular effects of shisha smoking: a cross-sectional study. *JRSM Open* 2014 May 6; 5(6): 2054270414531127. doi: 10.1177/2054270414531127. eCollection 2014.

[104] Selim GM, Fouad H, Ezzat S. Impact of shisha smoking on the extent of coronary artery disease in patients referred for coronary angiography. Anadolu *Kardiyol Derg* 2013 Nov; 13(7): 647-654. Doi: 10.5152/akd.2013.191. Epub 2013 Aug 28.

[105] Al Suwaidi J, Zubaid M, El-Menyar AA et al. Prevalence and outcome of cigarette and waterpipe smoking among patients with acute coronary syndrome in six Middle-Eastern countries. *Eur J Prev Cardiol* 2012 Feb; 19(1): 118-125. doi: 10.1177/1741826710393992. Epub 2011 Feb 22.

[106] World Health Organization: TobReg Advisory Note. Waterpipe tobacco smoking: health effects, research needs and recommended actions by regulators Geneva: *World Health Organization*; 2005.

[107] Mohammad Y, Kakah M, Mohammad Y. Chronic respiratory effect of narguileh smoking compared with cigarette smoking in women from the East Mediterranean region. *Int. J. Chron. Obstruct. Pulmon. Dis* 2008; 3: 405–414.

[108] Bachir R, Chaaya M. Maternal smoking: determinants and associated morbidity in two areas in Lebanon. *Maternal Child Health J.* 2008; 12: 298-307.

[109] Nuwayhid I, Yamout B, Azar G, Kambris M. Narghile (hubble-bubble) smoking, low birth weight, and other pregnancy outcomes. *Am. J. Epidemiol.* 1998; 148: 375-383.

[110] Dziuda MR, Grzybowski A. Cotinine – the biomarker of exposure to tobacco smoke. *Przegl. Lek* 1999; 56: 161–163.

[111] Oddoze C, Dubus JC, Badier M et al. Urinary cotinine and exposure to parental smoking in a population of children with asthma. *Clin. Chem.* 1999; 45: 505–509.

[112] Bacha ZA, Salameh P, Waked M. Saliva cotinine and exhaled carbon monoxide levels in natural environment waterpipe smokers. *Inhal Toxicol* 2007; 19: 771-777.

[113] Macaron C, Macaron Z, Maalouf MT et al. Urinary cotinine in narguila or hichi tobacco smokers. *J. Med. Liban* 1997; 45: 19–20.

[114] Maziak W, Rastam S, Shihadeh AL et al. Nicotine exposure in daily waterpipe smokers and its relation to puff topography. *Addict Behav* 2011; 36: 397-399.

[115] Neergaard J, Singh P, Job J, Montgomery S. Waterpipe smoking and nicotine exposure: a review of the current evidence. *Nicotine Tob. Res.* 2007; 9: 987-994.

[116] Vansickel AR, Shihadeh A, Eissenberg T. Waterpipe tobacco products: nicotine labelling versus nicotine delivery. *Tob Control* 2012 May; 21(3): 377-379. doi: 10.1136/tc.2010.042416. Epub 2011 Jun 2.

[117] Cobb CO, Shihadeh A, Weaver MF, Eissenberg T. Waterpipe tobacco smoking and cigarette smoking: a direct comparison of toxicant exposure and subjective effects. *Nicotine Tob. Res.* 2011; 13: 78-87.

[118] Lim BL, Lim GH, Seow E. Case of carbon monoxide poisoning after smoking shisha. *Int. J. Emerg. Med.* 2009; 2: 121–122.
[119] Maziak W, Rastam S, Ibrahim I et al. CO exposure, puff topography, and subjective effects in waterpipe tobacco smokers. *Nicotine Tob Res* 2009; 11: 806–811.
[120] Sajid KM, Akhter M, Malik GQ. Carbon monoxide fractions in cigarette and hookah (hubble bubble) smoke. *J. Pak. Med. Assoc*. 1993; 43: 179–182.
[121] Clarke SF, Stephens C, Farhan M et al. Multiple patients with carbon monoxide toxicity from water-pipe smoking. *Prehosp Disaster Med* 2012 Dec; 27(6): 612-614. Doi: 10.1017/S1049023X12001227.
[122] La Fauci G, Weiser G, Steiner IP, Shavit I. Carbon monoxide poisoning in narghile (water pipe) tobacco smokers. *CJEM* 2012 Jan; 14(1): 57-59.
[123] Bonadies N, Tichelli A, Rovó A. When water does not clear the smut from the smoke. BMJ Case Rep 2013 Oct 18; 2013. pii: bcr2013200665. doi: 10.1136/bcr-2013-200665.
[124] Saleh R, Shihadeh A. Elevated toxicant yields with narghile waterpipes smoked using a plastic hose. Food Chem Toxicol 2008; 46: 1461-1466.
[125] Zahran F, Yousef AA, Baig MH. A study of carboxyhaemoglobin levels of cigarette and sheesha smokers in Saudi Arabia. Am*. J. Public Health* 1982; 72: 722–724.
[126] Zahran FM, Ardawi MS, Al-Fayez SF. Carboxyhemoglobin concentrations in smokers of sheesha and cigarettes in Saudi Arabia. *Br. Med. J.* (Clin Res Ed). 1985; 291(6511): 1768–1770.
[127] Daher N, Saleh R, Jaroudi E et al. Comparison of carcinogen, carbon monoxide, and ultrafine particle emissions from narghile waterpipe and cigarette smoking: sidestream smoke measurements and assessment of second-hand smoke emission factors. *Atmos Environ* 2010; 44: 8-14.
[128] Fromme H, Dietrich S, Heitmann D et al. Indoor air contamination during a waterpipe (narghile) smoking session. *Food Chem. Toxicol.* 2009; 47: 1636-1641.
[129] Schweisfurth H. Beeinträchtigung der Atmungsorgane durch Aktiv- und Passivrauchen. *Atemw-Lungenkrkh* 2001; 27: 618-621.
[130] Maziak W, Rastam S, Ibrahim I et al. Waterpipe-associated particulate matter emissions. *Nicotine Tob. Res*. 2008; 10: 519-523.
[131] Cobb CO, Vansickel AR, Blank MD et al. Indoor air quality in Virginia waterpipe cafes. Tob Control 2013 Sep; 22(5): 338-343. doi: 10.1136/tobaccocontrol-2011-050350. Epub 2012 Mar 24.

[132] Zaidi SM, Moin O, Khan JA. Second-hand smoke in indoor hospitality venues in Pakistan. *Int. J. Tuberc. Lung. Dis.* 2011 Jul; 15(7): 972-977. doi: 10.5588/ijtld.10.0524.

[133] Bentur L, Hellou E, Goldbart A et al. Laboratory and clinical acute effects of active and passive indoor group water-pipe (narghile) smoking. *Chest* 2014 Apr; 145(4): 803-809. doi: 10.1378/chest.13-0960.

[134] She J, Yang P, Wang Y et al. Chinese Waterpipe Smoking and the Risk of Chronic Obstructive Pulmonary *Disease. Chest* 2014 Feb 20. doi: 10.1378/chest.13-1499. [Epub ahead of print].

[135] Harrabi I, Maaloul JM, Gaha R et al. Comparison of cigarette and waterpipe smoking among pupils in the urban area of Sousse, Tunisia. *Tunis .Med.* 2010; 88: 470-473.

[136] Riachy M, Rehayem C, Khoury C et al. Are narghile smokers different from cigarette smokers? *Rev. Mal. Respir.* 2008; 25: 313-318.

[137] Tamim H, Al-Sahab B, Akkary G et al. Cigarette and nargileh smoking practices among school students in Beirut, Lebanon. *Am. J. Health* Behav 2007; 31: 56-63.

[138] Chaaya M, El-Roueiheb Z, Chemaitelly H et al. Argileh smoking among university students: a new tobacco epidemic. *Nicotine Tob. Res.* 2004; 6: 457-463.

[139] Zoughaib SS, Adib SM, Jabbour J. Prevalence and determinants of water pipe or narghile use among students in Beirut's southern suburbs. *J. Med. Liban* 2004; 52: 142-148.

[140] Tamim H, Terro A, Kassem H et al. Tobacco use by university students, Lebanon, 2001. Addiction 2003; 98: 933-939.

[141] Baddoura R, Wehbeh-Chidiac C. Prevalence of tobacco use among the adult Lebanese population. *East Mediterr Health J.* 2001; 7: 819-828.

[142] Ward KD, Eissenberg T, Rastam S et al. The tobacco epidemic in Syria. *Tob. Control* 2006; 15 (suppl I): I 24-29.

[143] Maziak W, Ward KD, Mzayek F et al. Mapping the health and environmental situation in informal zones in Aleppo, Syria: report from the Aleppo household survey. *Intern. Arch. Occup. Environ Health* 2005; 78: 547-558.

[144] Maziak W, Eissenberg T, Rastam S et al. Beliefs and attitudes related to narghile (waterpipe) smoking among university students in Syria. *Ann. Epidemiol.* 2004; 14: 646-654.

[145] Maziak W, Hammal F, Rastam S et al. Characteristics of cigarette smoking and quitting among university students in *Syria. Prev. Med.* 2004; 39: 330-336.

[146] Al-Haddad N, Hamadeh RR. Smoking among secondary-school boys in Bahrain: prevalence and risk factors. *East Mediter Health J.* 2003; 9: 78-86.

[147] Behbehani NN, Hamadeh RR, Macklai NS. Knowledge of and attitudes towards tobacco control among smoking and non-smoking physicians in 2 Gulf Arab states. *Saudi Med. J.* 2004; 25: 585-591.

[148] Mandil A, Hussein A, Omer H et al. Characteristics and risk factors of tobacco consumption among University of Sharjah students 2005. *East Mediter Health J.* 2007; 13: 1449-1458.

[149] Amin TT, Amr MA, Zaza BO, Suleman W. Harm perception, attitudes and predictors of waterpipe (shisha) smoking among secondary school adolescents in Al-Hassa, Saudi Arabia. *Asian Pac. J. Cancer Prev.* 2010; 11: 293-301.

[150] Al-Mohamed HI, Amin TT. Pattern and prevalence of smoking among students at King Faisal University, Al Hassa, Saudi Arabia. *East Mediterr Health J.* 2010; 16: 56-64.

[151] Taha AZA: Prevalence of risk-taking behaviors. Bahrain Medical Bulletin 2007; 29: 1-10.

[152] Al-Turki YA. Smoking habits among medical students in Central Saudi Arabia. *Saudi Med. J.* 2006; 27: 700-703.

[153] Milaat WA, Al-Bar HS, Ghabrah TM et al. Preventive practices and non healthy behaviors among female university employees in Saudi Arabia. *Bahrain Medical Bulletin* 1999; 21: 75-79.

[154] AlSwuailem AS, AlShehri MK, Al-Sadhan S. Smoking among dental students at King Saud University: Consumption patterns and risk factors. *Saudi Dent. J.* 2014 Jul; 26(3): 88-95. doi: 10.1016/j.sdentj.2014.03.003. Epub 2014 May 14.

[155] Nisar N, Qadri MH, Fatima K, Perveen S. A community based study about knowledge and practices regarding tobacco consumption and passive smoking in Gadap Town, Karachi. *J. Pak. Med. Assoc.* 2007; 57: 186-188.

[156] Nisar N, Billoo N, Gadit AA et al. Pattern of tobacco consumption among adult women of low socioeconomic community Karachi, Pakistan. *J. Pak. Med. Assoc.* 2005; 55: 111-114.

[157] Alam SE. Prevalence and pattern of smoking in Pakistan. *J. Pak. Med. Assoc.* 1998; 48: 64-66.

[158] Rice VH, Weglicki LS, Templin T et al. Predictors of Arab American adolescent tobacco use. *Merrill-Palmer* 2006; 52: 327-342.

In: Nicotine Dependence
Editor: Carlton L. Owens
ISBN: 978-1-63482-642-6

Chapter 4

TREATMENT OF NICOTINE DEPENDENCE

***Rami Azmi Elshatarat, RN, CNS, MSc, PhD*[1,2,*], *Khaled A. Taha, MSc*[3], *Zyad T. Saleh, RN, MSc, PhD*[4], *and Mohammed A. Albalawi, PhD*[5]**

[1]Department of Physiological Nursing, School of Nursing, University of California, San Francisco, California, US
[2]College of Nursing Science, Al-Ghad International Colleges for Applied Medical Sciences, Al-Madinah Al-Munawarah, Kingdom of Saudi Arabia
[3]Pharmacy Department, Al-Ghad International Colleges for Applied Medical Sciences, Al-Madinah Al-Munawarah, Kingdom of Saudi Arabia
[4]Faculty of Nursing, University of Jordan, Amman, Jordan
[5]College of Medicine and College of Nursing Science, Taibah University, Al-Madinah Al-Munawarah, Kingdom of Saudi Arabia

ABSTRACT

Worldwide tobacco use, production, and consumption are continuing. Of approximately 1.2 billion smokers in the world, about half of them will die from diseases caused by smoking. Tobacco use is a major avoidable cause many diseases.

---

* Corresponding Author: Rami Azmi Elshatarat, RN, CNS, MSc, PhD. College of Nursing Science, Al-Ghad International Colleges for Applied Medical Sciences, Al-Madinah Al-Munawarah, Kingdom of Saudi Arabia; Cell phone: +966 509 141 940; Phone: +966 4 8224841. Ex: 115; Fax: +966 4 8224605; Email: elshatrat@hotmail.com

Despite of recommendations to stop smoking, smokers with life-threatening medical disorders are unable to quit smoking. Approximately, 80% of American smokers who made attempt to quit smoking independently by themselves, relapsed within the first month of abstinence. These quitting attempt failures may relate to nicotine dependence and/or low self-efficacy to quit.

In 1988, the Surgeon General's report on the health consequences of smoking definitively concluded that all tobacco forms were powerfully "addicting drugs." Nicotine is a pharmacologic agent that has physiological and psychological effects on the human body. Nicotine has effects on mood and cognitive function that contribute to tobacco addiction. Smokers report positive effects such as pleasure and relaxation, which triggers them to smoke.

According to treating tobacco use and nicotine dependence guidelines, tobacco dependence is recognized as a chronic disease and requires serious interventions. Five nicotine replacement therapies (NRTs: nicotine inhalers, lozenges, gum, nasal sprays, and patches) and two non-NRTs (bupropion SR, and varenicline) are the first line to treat nicotine dependence. Nicotine replacement medications are recommended to maintain abstinence from tobacco, avoid nicotine withdrawal symptoms and prevent relapse. Combinations of counseling and medication are recommended to achieve effective treatment and increase long-term smoking cessation rates.

The goal of this review chapter is to describe the nicotine dependence, discuss the neurobiology of nicotine dependence, which is important for understanding tobacco addiction and its treatment, and to describe the pharmacological and non pharmacological treatments of nicotine dependence.

Keywords: neurobiology of nicotine dependence; treatment of nicotine dependence; smoking cessation; theories of nicotine dependence

INTRODUCTION

Worldwide tobacco use, production, and consumption are continuing (U.S. Department of Health and Human Services [USDHHS], 2014). Of approximately 1.2 billion smokers in the world, about half of them will die from diseases caused by smoking (Mackay, Ericksen & Shafey, 2006). Tobacco use has been recognized as a chief avoidable cause of multiple cancers, cardiovascular disease, stroke, complications of pregnancy and chronic obstructive pulmonary disease (USDHHS, 2004, 2014). Smoking cessation reduces the morbidity and mortality associated with such diseases (USDHHS, 2004, 2014).

Approximately, one third of the adult male population globally smokes. Smoking related-diseases kill 1 in 10 adults globally. Every eight seconds, someone dies from tobacco use. If current trends continue, smoking is estimated to kill one in 6 people worldwide (Centers for Disease Control and Prevention [CDC], 2013; Eriksen, Mackay & Ross, 2012; USDHHS, 2014). Smoking prevalence is on the rise in the developing world, whereas, it is decreasing in developed nations. In the developing world, tobacco consumption is rising by 3.5% per year (World Health Organization [WHO], 2002; WHO. 2013).

Tobacco use is the most preventable cause of death worldwide. It causes more than 435,000 deaths each year in the United States (CDC, 2007). Despite the harmful consequences of tobacco use and the population's awareness of these health problems, approximately 1,200 children and adolescents become new cigarette smokers each day (Frioe et al., 2008; Substance Abuse and Mental Health Services Administration, 2007; USDHHS, 2014). Even with recommendations to stop smoking, smokers with life-threatening medical problems such as acute coronary syndrome or asthma are unable to quit smoking (USDHHS, 2014). In 2005, only 4 to 7 percent of 19 million American adults who attempted to quit smoking were successful (CDC, 2006). In the United States, unfortunately, 80% of smokers who made attempt to quit by themselves, relapsed within the first month of abstinence and only 3% were remain abstinent at 6 months (Benowitz, 2008a). These quitting attempt failures may relate to nicotine dependence and/or insufficient motivating factors such as the individual's confidence in their ability (self-efficacy) to quit (Fiore et al., 2008). Combinations of counseling and medication are effective and recommended for treating tobacco dependence in all individuals who want to quit (Fiore et al., 2008; USDHHS, 2014).

Many smokers face great challenges when quitting smoking. These attempts to quit fail due to the addictive properties of nicotine. Nicotine is present in all forms of tobacco. So, the understanding of health care providers about the nature of tobacco addiction will help health organizations, community groups and government agencies to successfully run tobacco free programs (US Department of Health and Human Services [USDHHS], 1988 Fiore et al., 2008; USDHHS, 2014).

In addition to combination of pharmacological therapy and counseling, behavioral therapy through improving smokers' self-efficacy is recommended to treat nicotine dependence (Fiore et al., 2008; USDHHS, 2014). Self-efficacy theory has been studied to assess its effects on health promotion in various health activities such as smoking cessation, cancer prevention and changing behavior related to drug dependence (Badr & Moody, 2005; Smith & Liehr, 2003). Many studies have found that self-efficacy has an important role in improvement of individual performance and change in their adverse behaviors (Bandura, 1999; Smith & Liehr, 2003). The self-efficacy concept has been widely accepted by psychologists for dealing with a variety of human behaviors (Lee, 1989). Moreover, Bandura's self-efficacy theory is used in nursing studies as well as for therapeutic intervention to improve individual behavior (Lee, 1989; Smith & Liehr, 2003). Self-efficacy is considered to have an important role in influencing and changing many adverse behaviors including smoking cessation (Badr & Moody, 2005).

The goal of this chapter is to discuss the neurobiology of nicotine dependence, which is important for understanding tobacco addiction and its treatment. Specifically, this chapter will discuss: nicotine dependence, the mechanism of nicotine addiction, the neurobiology of nicotine dependence, and therapeutic approaches to nicotine addiction treatment.

DEFINITION OF NICOTINE DEPENDENCE

A drug is defined as "a substance that, when taken into the living organism, may modify one or more of its functions" (Edwards, Arif & Hadgson, 1981, p.6). Drug dependence is defined as a "socio-psycho-biological syndrome manifested by a behavioral pattern in which the use of a given psychoactive drug (or class of drugs) is given a much higher priority over other behaviors which once had significantly greater value" (Edwards, Arif & Hadgson, 1981, p.19). The USDHHS (1988) described nicotine as a psychoactive drug with actions that reinforce the tobacco use. Edwards, Arif and Hadgson (1981) stated that drug dependence includes psychic dependence, which is defined as "a feeling of satisfaction and a psychic drive that requires periodic or continuous administration of the drug to produce pleasure or to avoid discomfort" (p.4). Also, they defined physical dependence as

> an adaptive state that manifests itself by intensive physical disturbances when the administration of the drug is suspended ... These disturbances, i.e. the withdrawal or abstinence syndromes, are made up of specific arrays of symptoms and signs of psychic and physical nature that

are characteristic for each drug type (Edwards, Arif & Hadgson, 1981, p.4).

The term "nicotine addiction" is not new. In 1964, the Report of the Advisory Committee to the Surgeon General used the term "habituating" to refer to tobacco use (USDHHS, 1964; 1988). After the research was complete, the Report of the Surgeon General in 1979, said that smoking is "the prototypical substance-abuse dependency" (USDHHS, 1979). In 1988, the Surgeon General's report on the health consequences of smoking definitively concluded that cigarettes and other tobacco forms were powerfully "addicting drugs" (USDHHS, 1988). The terms "dependence" and "addiction" are used interchangeably in the literature. These terms describe the same neurochemical and psychological processes involved in drug use (Benowitz, 1999; USDHHS, 1988; West, 2006). The term "nicotine dependence" is used more commonly than "nicotine addiction" (West, 2006).

Table 1. Surgeon General's Report: Criteria or Drug Dependence

Primary criteria:
- Highly controlled or compulsive use
- Psychoactive effects
- Drug-reinforced behavior

Additional criteria:
- Addictive behavior often involves:
  - Stereotypic patterns of use
  - Use despite harmful effects
  - Relapse following abstinence
  - Recurrent drug cravings
- Dependence-producing drugs often produce:
  - Tolerance
  - Physical dependence
  - Pleasant (euphoriant) effects

The criteria from: U.S. Department of Health and Human Services. The health consequences of smoking: Nicotine addiction, a report of the Surgeon General (DHHS Publication No. CDC 88-8406). Washington, DC: U.S. Government Printing Office. 1988.

The Surgeon General in 1988 developed criteria for drug dependence which included primary and additional criteria (Table 1). Primary criteria define drug dependence, and describe nicotine's psychoactive effects.

Secondary criteria are used to characterize drug dependence such as the development of drug tolerance, drug cravings, use despite harmful effects, and probability of relapse after cessation (USDHHS, 1988, p 7).

EVIDENCE THAT NICOTINE USE IS ADDICTING

Understanding the pharmacology of nicotine and the neurobiology of nicotine dependence is important for health care providers to help tobacco users to quit smoking and prevent relapse. Since 1964, the term "drug habit" was dropped and replaced by drug dependence. The term tobacco addiction began to be used in the literature in the early 1900s (USDHHS, 1988). The scientific evidences support the idea of nicotine as an addictive drug. This evidence is inferred from experimental studies (e.g., studies that examine the effect of nicotine on mice), clinical studies, and from the theoretical background of the characteristics and actions of nicotine and the diagnostic criteria of addiction (USDHHS, 1988). For instance, the USDHHS (1988) concluded that nicotine has pharmacological effects and produces addiction similar to other addictive drugs such as heroin and cocaine (National Institute on Drug Abuse, 2009; USDHHS, 1988; 2010; 2014). Scientific evidence demonstrates that nicotine is a psychoactive drug. Nicotine affects all of the human body, particularly the brain. Repeated exposure to nicotine develops neuroadaptation that leads to nicotine tolerance and addiction (Benowitz, 1999; 2008a; 2008b; USDHHS, 1988; 2014).

There is much evidence that explains the role of nicotine in sustaining tobacco use. The addictive properties of smoking in spite of smoking-related diseases, is one of the major explanation why smokers are unable to quit smoking (Benowitz, 1999; 2008a; 2008b; USDHHS, 1988; 2014). Benowitz (1999; 2008b) described the evidence for tobacco addiction: First, tobacco users who smoke are unable to quit easily. Tobacco users who have smoking-related diseases such as cardiovascular disease and cancer fail to quit smoking even after multiple serious quit attempts. On average, most tobacco users who try to stop smoking need four attempts to do so. Approximately 97% of smokers' unaided quit attempts fail to achieve long-term abstinence. Half of smokers are unable to stop smoking, even though they know that continuing smoking may kill them. Secondly, nicotine is a pharmacologic agent that has physiological and psychological effects on the human body. Nicotine has effects on mood and cognitive function that contribute to tobacco addiction. Smokers report positive effects such as pleasure, relaxation and improved performance, which trigger them to smoke. Finally, maintaining abstinence from tobacco, avoiding nicotine withdrawal symptoms, craving nicotine and preventing relapse can be facilitated by treatment with nicotine replacement medications (Benowitz, 1999; 2008b).

In addition, the nicotine dependence meets many criteria used to diagnose drug addiction. For example, nicotine use meets all points of the Surgeon General's primary criteria for drug addiction (Table 1) (USDHHS, 1988, p 7). Moreover, according to the DSM-IV criteria, tobacco use is considered a drug addiction when at least three of seven criteria are met within the same 12-month period (American Psychiatric Association, 1994, p 181)

Benowitz (1999) noted that three of these criteria are relevant to tobacco use: nicotine is often taken in larger amounts or over a longer period than was intended, unsuccessful efforts to cut down or control tobacco use, and continuing to use nicotine despite the development of physical or psychological problems that are likely to have been caused or exacerbated by tobacco. Cohen et al. (2003) stated that the DSM-IV criteria for the development of tolerance to nicotine and the experience of withdrawal are also relevant to tobacco use (Cohen et al., 2003).

ETIOLOGICAL RISK FACTORS FOR NICOTINE DEPENDENCE

Tobacco addiction is due to the psychological and neurobiological effects of nicotine, which is found in all tobacco products (Cohen, et al., 2003; USDHHS, 1988; 2014). Repeated tobacco use in some specific situations (certain environments, emotional states, and psychosocial stimuli such as depressed mood and anxiety) automatically triggers smokers' desire for more tobacco use (Cohen, et al., 2003). Hatsukami et al. (2008) argued that in addition to the physiological contribution of nicotine dependence, there are many other factors that play an important role in nicotine addiction (Hatsukami, Stead & Gupta, 2008). These factors include the environmental culture: availability of tobacco products, bans on tobacco use, cost of tobacco, social acceptability, and modeling such as parental or peer smoking. They also include the characteristics of the individual: e.g., genes, co-morbid psychiatric disorders, and personality features (Hatsukami et al., 2008). Data show there is variation in genes such as cholinergic nicotine receptor genes, dopamine pathway (dopamine D2 and D4) receptor genes and dopamine transport genes

(Hatsukami et al., 2008). In addition to neurobiological and psychosocial factors, the genetic factor is a risk factor for nicotine dependence (Cohen, et al., 2003; Hatsukami et al., 2008). Nicotine dependence also varies across gender. The proportion of tobacco users among men is much higher than among women in some countries such as Japan and Greece, but not in others, such as the United States and the United Kingdom (Cohen, et al., 2003).

NEUROBIOLOGY OF NICOTINE DEPENDENCE

Tobacco addiction is a complex neuro-psycho-pharmacological disorder. Addiction is a process involving multiple molecular and brain components (Benowitz, 1999; 2008a; Rose, 2007). One of the major challenges that face neuroscientists is to piece together the chemical and molecular alterations of the brain in response to addictive drugs (Kelley, 2002).

There are several chemical compounds in tobacco; nicotine is the main source of tobacco addiction (Benowitz, 2008a; Sohn, et al., 2003; USDHHS, 1988; 2014). It is the most pharmacologically active substance in tobacco. All tobacco forms contain nicotine (USDHHS, 1988, 2014). Nicotine is a weak base, which mean that half of nicotine is ionized and the other half is non-ionized (USDHHS, 1988, 2014). When a person smokes, nicotine is absorbed from the tobacco and is carried into the lungs. Nicotine is rapidly absorbed from aveolar space into pulmonary venous circulation because of the large surface area of the alveoli and relatively high pH (7.4) of the lung fluid (Benowitz, 2008a; USDHHS, 1988). Then nicotine is carried into the pulmonary venous circulation. Then, it reaches the arterial circulation. Absorbed nicotine diffuses readily into brain tissue and binds to nicotinic acetylcholine receptors (nAChR) (Benowitz, 2008a). The rapid delivery of nicotine to the brain increases its psychoactive rewards, which is associated with drug dependence (Benowitz, 2008a; Sohn, et al., 2003).

Nicotine increases activity in the prefrontal cortex, thalamus, corticobasal ganglia-thalamic brain circuits and visual system. When nicotine binds to nicotine receptors, changes in the states of the channels will occur. These changes include the resting (an activated) state (channel open) and two desensitization states (channel closed). Both transition and persistent desensitization states of nicotine receptors lead to nicotine tolerance (Benowitz, 2008a, 2008b).

Nicotine acts presynaptically to facilitate release of many brain neurotransmitters, including norepinephrine, acetylcholine, serotonin, γaminobutyric acid (GABA), glutamate, endorphins and the most important neurotransmitter, dopamine. Dopamine produces a pleasurable sensation, which leads to the reinforcing effects of nicotine and other drugs of abuse (Benowitz, 1999; 2008a, 2008b). These neurotransmitters mediate various addiction behaviors. These neurotransmitters lead to reduction of anxiety, pleasure, arousal, cognitive enhancement, mood modulation, performance enhancement, and analgesic and weight-loss effects (Figure 1) (Benowitz, 1999; 2008a).

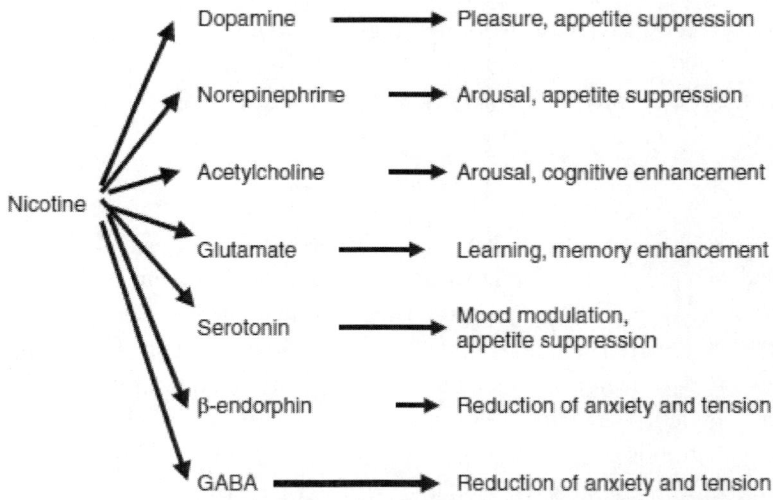

The figure from: Benowitz, N. L. (2008). Clinical pharmacology of nicotine: Implications for understanding, preventing, and treating tobacco addiction. *Clinical Pharmacology & Therapeutics*, 83(4), 531-541.

Figure 1. Effect of Nicotine Use on Nicotinic Cholinergic Receptor and Smokers' behavior.

Repeated exposure to nicotine contributes to neuroadaptation, from which the phenomenon of tolerance develops (Benowitz, 2008a; Hatsukami, et al., 2008; Sohn, et al., 2003). Tolerance is the state which occurs after repeating doses of a drug, and then given a dose of the drug will not produce the same effect. The dose produces less of an effect than before, and increasing doses are required (Benowitz, 1988). Neuroadaptation occurs to attain homoeostasis resulting from the activation of the nAChRs sites and increased concentration of neurotransmitters (Hatsukami, et al., 2008). Chronic nicotine exposure leads to upregulatation of some subtypes of nAChRs, an increase in numbers of nAChRs, and produces desensitized (inactivated) nicotine receptors, which are critical for the development of nicotine tolerance (Benowitz, 2008a; Hatsukami, et al., 2008). In addition, chronic exposure to nicotine reduces brain monoamine oxidase A and B activity. This results in increased monoaminergic neurotransmitter levels such as dopamine and norepinephrine in the synapses, which plays a major role in augmenting the effects of nicotine and contributing to nicotine dependence (Brody, 2006).

Craving and withdrawal symptoms occur during periods of nicotine abstinence, i.e. night time sleep, when the desensitized α4β2 nAChRs become unoccupied. This explains the daily smoking behavior aimed to maintain the body nearly saturated with nicotine, thus occupying (desensitized) brain nAChRs. Smokers then start smoking soon after they wake up. When smoking, nicotine binds with nicotine receptors, craving and withdrawal are avoided and alleviated through maintaining nAChR desensitization (Benowitz, 1999; 2008a).

Quitting smoking decreases the release of neurotransmitters, particularly dopamine, causing the body to develop a homeostatic response. This causes withdrawal symptoms on abstinence from nicotine (Hatsukami, et al., 2008; Benowitz, 2008a). For example, in the state of dopamine deficiency, smokers are unable to experience pleasure (Benowitz, 2008a). Nicotine withdrawal symptoms include negative affect such as irritability, restlessness, frustration or anger, anxiety, difficulty concentrating, depressed mood, insomnia, increased appetite or weight gain, and decreased heart rate (Benowitz, 2008a; Hatsukami, et al., 2008; Sohn, et al., 2003). Other possible symptoms can include constipation, cough, dizziness, increased dreaming and mouth sores (Hughes, 2007). The withdrawal symptoms cause impairment in social, occupational, or other areas of functioning (Benowitz, 2008a).

Nicotine withdrawal symptoms begin within 1 to 2 days of cessation, peak during the first week, and gradually decrease to pre-cessation levels within approximately 4 weeks (Benowitz, 2008b; Sohn, et al., 2003). Other studies have shown that nicotine withdrawal symptoms might persist for several months after quitting smoking (Piasecki, Fiore, McCarthy, et al., 2002). Nicotine withdrawal symptoms might increase over the 6 months after quitting, and then decrease or become sustained (Hughes, Higgins & Hatsukami, 1990).

Nicotine dependence is produced by a combination of positive reinforcements such as enhanced mood and experience of pleasure and by avoidance of negative reinforcements of nicotine such as relief of withdrawal symptoms. In addition to the direct pharmacologic actions of nicotine, environmental conditioning is involved in the development of nicotine dependence. Environmental conditioning factors such as the smell of a cigarette, a meal, cup of coffee, talking on the phone, or drinking with friends are common triggers for smoking (Benowitz, 2008a, 2008b; Hatsukami, et al., 2008). All drug-taking behavior is learned, a result of conditioning. Conditioning is a major risk factor for relapse to tobacco use after smoking cessation (Benowitz, 2008a, 2008b).

THEORIES RELATED TO DRUG ADDICTION

Tobacco addiction is a complex neuropsychopharmacological disorder. Addiction is a process of multiple molecular and brain components (Rose, 2007). However, no comprehensive theory exists that includes all the dimensions of drug addiction. Many theories describe important elements for understanding drug addiction (West, 2006).

The Incentive Sensitization Theory of Addiction

The incentive sensitization theory of addiction was developed 15 years ago. Addiction results from drug-induced sensitization in the brain's mesocorticolimbic systems. This sensitization can cause pathological incentive motivation ('wanting') for drugs (Robinson & Berridge, 2008). Repetitive drug use leads to alteration in a number of psychological processes associated with various symptoms of addiction. Sensitized incentive salience resulting from a change in brain circuits is manifested by behavioral implicit (unconscious wanting) or explicit (conscious craving) processes depending on the circumstances. Incentive salience is associated with learning mechanisms but this learning per se is not enough for the pathological motivation for drug dependence (Robinson & Berridge, 2008).

Robinson and Berridge (1993) listed the four major tenets of incentive-sensitization: (1) addictive drugs alter brain organization; (2) alteration of brain systems is involved in the process of incentive motivation and reward; (3) neuroadaptation for addiction renders these brain reward systems hypersensitive ("sensitized") to drugs and drug-associated stimuli; and (4) the sensitized brain systems do not mediate the pleasurable or euphoric effects of drugs (drug "liking"). Instead, they mediate a subcomponent of reward we have termed incentive salience (drug “wanting”) (Robinson & Berridge, 1993).

Self-medication Model of Addiction

The self-medication hypothesis proposes that the reason for drug use is to cope with an underlying psychological or social disorder by means of self-medication (Robinson & Berridge, 2000). This model provides some rationale for the drug user to choose drugs as a method of treating their underlying psychological problems and to help them control their aggressive impulses. For example, one of the major motivating factors for smoking is to relieve stress (West, 2006).

The Disease Model of Addiction

This model argues that addiction is the result of pathological changes in the brain. For example, repeat exposure to nicotine over the time causes abnormalities in structure and function of the central nervous system that results in overpowering urges (West, 2006).

Robinson and Berridge (2000) stated that the disease model and the self-medication hypothesis are the most effective theories of drug addiction. The disease model proposes that the drug addict is sick and suffers from a disease.

The Dopamine Theory of Addiction

This theory proposes that the action of dopamine on receptors in the nucleus accumbens has effective action producing addiction. Addictive drugs release the neurotransmitter dopamine in the brain, particularly in the nucleus accumbens. The nucleus accumbens increases of dopamine concentration that corresponds to the development of addiction. Nicotine can block the inhibitory control of GABA in the ventral tegmental area (VTA) leading to firing of neurons in the nucleus accumbens. This process pathway is reinforces and cues drug-seeking behavior (Benowitz, 2008a; Robinson & Berridge, 2008).

TOBACCO USE PREVENTION AND CONTROL OF NICOTINE DEPENDENCE

Cigarette smoking has been universally accepted as a major cause of serious disease and death all over the world (USDHHS 2010; 2014). Recent research demonstrates the hazards of smoking and its contribution to many diseases. The intensity of nicotine dependence is dependent upon various factors such as age of starting smoking, form of tobacco use, cigarette consumption per day, and duration of tobacco use (Ezzati & Lopez, 2003, Fiore et al. 2008; USDHHS 2010; 2014). There are many risk factors for nicotine addiction such as psychosocial, environmental, and economic factors for the smoker and society at large. There are many risk factors that can be controlled. Knowledge of the etiological risk factors for nicotine dependence and the neurobiology of nicotine are basic requirements for understanding prevention, control and treatment of nicotine dependence (Benowitz, 2008a, 2008b; Hatsukami, et al., 2008, USDHHS, 2014).

Health institutions and government agencies collaborate to control tobacco use and to create communities free of tobacco (USDHHS, 2014). Smoking cessation programs and policies are established to protect people from the hazards of smoking. These policies vary from country to country. Some have used the WHO Tobacco Convention to legally counter the tobacco industry's aggressive efforts to get new smokers initiated (WHO, 2013).

Many countries commonly use tobacco control regulations (WHO, 2013). These regulations include: raising cigarettes taxes, restricting smoking in public places and on public transportation, prohibiting tobacco advertising, restricting the sale of tobacco products to minors, and putting warning labels about the health hazards of nicotine addiction on tobacco packets (Hall, Gartner & Carter, 2008; USDHHS, 2000, WHO, 2009). There are penalties for persons who violate these regulations. In addition to these regulations, many smoking cessation programs have been initiated. These programs use media campaigns such as television ads, radio, official newspapers, billboards, and publications aiming to fight tobacco use, promote health awareness of the hazards of smoking and nicotine dependence, and to provide advices on how quit smoking (USDHHS, 2000, 2014; WHO, 2009, 2013).

TREATMENT OF TOBACCO USE AND NICOTINE DEPENDENCE

The "5 A's" Model for Treating Tobacco Use and Dependence

Fiore, et al. (2008) described the "5 As" model for treating tobacco use and dependence (Ask-screening all patients for tobacco use, Advice to quit, Assess willingness to quit, Assist with quitting and Arrange follow up) (Figure 2) (Fiore, et al., 2008). The treatment strategies for this model are as follows: the clinician should assist the tobacco users who are willing to quit smoking by offering medication and providing counseling or additional treatment based on the tobacco users' needs and arrange for follow up contacts to prevent relapse. For tobacco users who are unwilling to make a quit attempt, clinicians should motivate and advise them to quit smoking in the future (Fiore, et al., 2008; USDHHS, 2014).

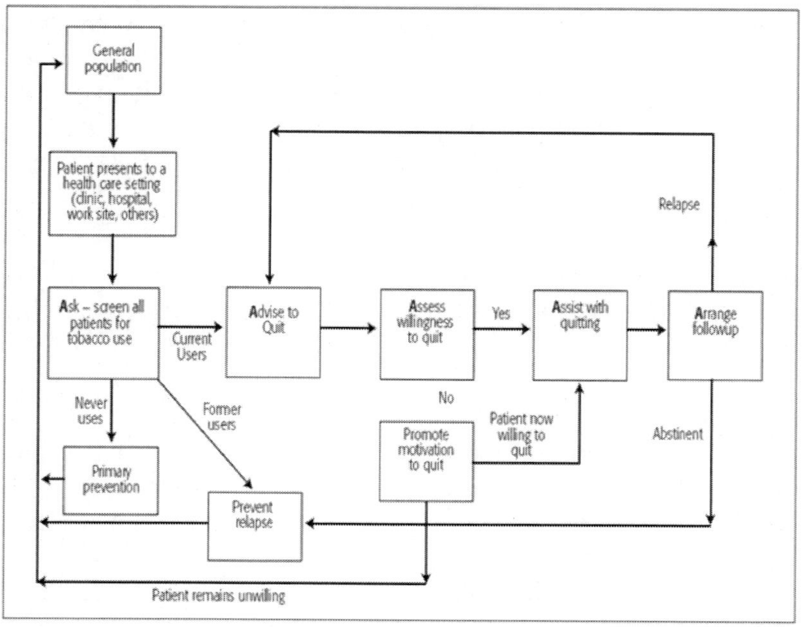

The figure from: Fiore M. C., Jaen C. R., Baker T. B., Bailey, W. C., Benowitz, N. L., Curry,N. J. et al. (2008). *Treating tobacco use and dependence: 2008 update.* Rockville, MD: US Department of Health and Human Services, Public Health Service.

Figure 2. The "5 A's" Model for Treating Tobacco Use and Dependence.

Table 3. Pharmacotherapy treatment of Nicotine Dependence

	Dose	Instructions for use	Most frequent side effects (at all doses)	Precautions (at all doses)
First-line therapies				
Nicotine-replacement therapies				Cardiac disease*, recent myocardial infarction, serious arrhythmias, unstable angina pectoris; peptic ulcers; insulin-dependent diabetes; uncontrolled hypertension; drug treatment for depression or asthma; pregnancy or lactation; hyperthyroidism, phaeochromocytoma or type 1 diabetes mellitus
Nicotine patch	21 mg,14 mg, 7 mg; 15 mg, 10 mg, 5 mg; 22 mg, 11 mg	24 h application: 21 mg for 4–6 weeks, 14 mg for 2 weeks, 7 mg for 2 weeks; 16 h application: 15 mg 6 weeks, 10 mg 2 weeks, 5 mg 2 weeks; 22 mg and 11 mg (light smokers) for 8 weeks	Local skin reaction, sleep disruption	Allergy to tape or dermatological condition
Nicotine gum	2 mg (<25 cpd); 4 mg (≥25 cpd)	1 piece every 1–2 h for 6 weeks, every 2–4 hours for 3 weeks, and every 4–8 hours for 3 weeks. Maximum=24 pieces	Jaw ache, mouth soreness, dyspepsia, hiccups	Dental conditions exacerbated by chewing gum; history of oesophageal, oral, or pharyngeal inflammation

Table 3. (Continued)

	Dose	Instructions for use	Most frequent side effects (at all doses)	Precautions (at all doses)
Nicotine inhaler	10 mg cartridge (4 mg delivered	6–16 cartridges per day for 12 weeks, gradual reduction for final 6–12 weeks	Local irritation of mouth and throat, coughing and rhinitis	Allergy to menthol, bronchospastic disease
Nicotine nasal spray	0.5 mg in each nostril	1–2 doses per h, 12–26 weeks of treatment. Maximum=40 doses per day	Nasal and airway irritation	Severe reactive airway disease, chronic nasal disorder
Nicotine lozenge	2 mg (≥30 min to first cigarette in morning); 4 mg (<30 min to first cigarette in morning)	1 lozenge every 1–2 h for 6 weeks, every 2–4 h for 3 weeks, and every 4–8 hours for 3 weeks. Maximum=20 per day	Hiccups, mouth or throat irritation, heartburn or dyspepsia	
Nicotine microtab	2 mg	1 (<20 cigarettes per day) – 2 (≥20 cigarettes per day) per h for 12 weeks, followed by gradual reduction. Maximum=40 per day		
Non-nicotine products				
Bupropion	150 mg twice a day†	1–2 weeks before quitting, 150 mg every morning for 3 days then 150 mg twice a day, 7–12 weeks of treatment, may consider extending treatment to 6 months (150 mg per day)	Dry mouth, insomnia	Seizure disorder or predisposition towards seizure disorder, concomitant use of wellbutrin, diagnosis of bulimia or anorexia nervosa, concomitant or recent use of monoamine oxidase inhibitors

	Dose	Instructions for use	Most frequent side effects (at all doses)	Precautions (at all doses)
Varenicline	1 mg twice a day	1 week before quitting, 0.5 mg once a day for days 1–3, 0.5 mg twice per day for days 4–7, 1 mg twice per day for 12 weeks, and an option of additional 12 weeks for quitters	Nausea/vomiting, sleep disturbance, flatulence	Severe renal impairment, pregnancy, can impair ability to drive or to operate heavy machinery
Second-line therapies				
Clonidine	Oral=0.15–0.75 mg per day; Patch=0.1–0.2 mg per day	Start on quit date or up to 3 days before quit date, initial dose 0.1 mg twice a day orally or 0.1 mg per day transdermally, increasing by 0.1 mg per day per week, 3–10 weeks of treatment, down titrate dose over 2–4 days	Dry mouth, drowsiness, dizziness, sedation, constipation, lower blood pressure, rebound hypertension	Pregnancy
Nortriptyline	75–150 mg per day	Initial dose 25 mg per day and up titrate to 75–150 mg per day 10–28 days before quit date, 12 weeks of treatment, may consider treatment for up to 6 months	Dry mouth, sedation, blurred vision, urinary retention, light headedness, shaky hands, cardiotoxicity with overdose	Cardiovascular disease, pregnancy, monoamineoxidase inhibitors, may impair ability to drive or to operate heavy machinery

cpd=cigarettes per day. *Treatment with nicotine patches has been examined with smokers who have cardiovascular disease and the drugs have been found to be safe. † No significant differences in cessation rates were recorded between the recommended 300 mg dose and the 150 mg dose (OR 1.07, 95% CI 0.987-1.32).

The pahrmacotherapies from: Hatsukami, D. K., Stead, L. F., & Gupta, P. C. (2008). Tobacco addiction. *The Lancet*, 371, 2027-2038.

There are many effective medications available for tobacco dependence. These pharmacologic treatments are approved by the US Food and Drug Administration (FDA) (Benowitz, 2008a; Fiore, et al., 2008; USDHHS, 2014). Fiore et al. (2008) listed seven first-line medications (five nicotine replacement therapies (NRT) and two non-NRT). These medications include: Nicotine gum, nicotine inhaler, nicotine lozenges, nicotine nasal spray, nicotine patch, Bupropion SR, and Varenicline (Table 3)

Pharmacotherapies

Nicotine Replacement Therapies (NRT) can be used instead of tobacco products to reduce nicotine intake and its addictive effects gradually which enhance smoking cessation (Benowitz, 2008a, 2008b; Fiore, et al., 2008; Hatsukami, et al., 2008; USDHHS, 2014).

NRT acts to relieve craving and withdrawal symptoms when the tobacco user quits. NRT reduces the amount and speed of nicotine delivery, desensitize of nicotinic receptors, so it reduces reinforcement from tobacco use and reduces tobacco toxicity (Benowitz, 1999; 2008a; Fiore, et al., 2008; Hatsukami, et al., 2008). Moreover, desensitization of nicotine receptors leads to reduce the effect of nicotine from tobacco products on the current tobacco users. So if a person back to smoke while taking NRT, the cigarette may be less satisfying. There is evidence that NRT can be effective in both men and women, but using NRT is contraindicated during pregnancy (Fiore et al., 2008)

Non-Nicotine Replacement Therapies (Non-NRT) include agents such as Bupropion, Nortriptyline, and Varenicline. Bupropion is an antidepressant drug. It functions as a non-competitive antagonist of the nicotinic acetylcholine receptor. It blocks the reuptake of dopamine and noradrenaline. Nortriptyline is a reuptake inhibitor for noradrenaline and serotonin (Hatsukami, et al., 2008).

Recently, a new drug has been released, Varenicline, a specific $\alpha 4\ \beta 2$ nicotinic receptor partial agonist which acts on dopamine stimulation and blocks nicotine receptors. $\alpha 4\ \beta 2$ nAChR partial agonists aim to reduce craving and withdrawal symptoms. It is an effective pharmacotherapeutic tool for quitting tobacco use and reducing risk for relapse (Benowitz, 2008a; Foulds, 2006; Hatsukami, et al., 2008).

Clonidine is a sympatholytic medication ($\alpha 2$ adrenergic agonist) that has been used to treat high blood pressure, attention-deficit, hyperactivity disorder, anxiety disorders, migraine, and withdrawal symptoms from alcohol, opioids, or smoking. Clonidine helps in calming and reduces anxiety. Therefore, it became useful to tobacco users who have experience a high degree of anxiety when they trying to quit smoking and it is considered as an effective smoking cessation treatment (Benowitz, 2008a; Fiore et al., 2008; USDHHS, 2014).

Many trials are in progress examining a *nicotine vaccine*, which stimulates the immune system to develop antinicotine antibodies. The results of these trials showed that the nicotine vaccine reduces the distribution of nicotine to the brain in rats (Hatsukami, et al., 2008). A second clinical trial is ongoing on testing a selective Cannabinoid CB1 receptor antagonist. This drug can block nicotine-induced dopamine release in the nucleus accumbens and reduces nicotine-seeking behavior (Hatsukami, et al., 2008).

Counseling

Counseling is an important non-pharmacological intervention for quitting smoking. Teaching smokers about the benefits of smoking cessation, problem solving skills and providing cognitive behavioral therapy, and social support as part of treatment can motivate tobacco users to quit smoking (Benowitz, 2008a; Fiore et al., 2008; Hatsukami, et al., 2008; USDHHS, 1988, 2014). The established benefits of counseling can be further enhanced by referring tobacco users to specialists in smoking cessation or to telephone quit lines when smoking cessation specialists are unavailable (Hatsukami, et al., 2008).

Combination of Pharmacological Treatments and Counseling

In United States, the USDHHS collaborates with federal agencies and nonprofit organizations in publishing and updating clinical practice guidelines. The USDHHS has published more than 25 reports about the treatment of tobacco use, management of nicotine withdrawal symptoms and prevention of relapse (USDHHS, 1988, 2014). These treatment guidelines have concluded that tobacco dependence requires serious interventions to quit tobacco use (Fiore, et al., 2008; USDHHS, 1988, 2014). According to the United Kingdom and US Public Health Service guidelines, all smokers should be considered for pharmacotherapy. It can improve quit rates by 2- to 3-fold (Fiore, et al., 2008; Hatsukami, et al., 2008; USDHHS, 2014). Many studies reported that combination of pharmacological drugs were more effective to quit tobacco use

(Fiore, et al., 2008; Hatsukami, et al., 2008; Silagy et al., 2004). For example, combination of nortriptyline and nicotine patches are more likely to improve smoking cessation rate than use nicotine patches alone (23% vs 10% at 6 months) (Hatsukami, et al., 2008; Prochazka et al., 2004). Moreover, combinations of counseling and medication are recommended to achieve effective treatment and increase long-term smoking cessation rates (Fiore, et al., 2008; USDHHS, 2014).

TOBACCO USE AND SELF-EFFICACY

Tobacco users face difficult challenges quitting smoking and maintaining abstinence. Despite difficulties in quitting smoking, 90% of ex-smokers quit smoking by themselves (Bandura, 1999). Moreover, smokers may need to repeat attempts to successfully quit. Many smokers need help to do so (Bandura, 1999). Environmental factors and negative affect are major motivators for the resumption of tobacco use among individuals who had quit. Exposure to environmental situations may induce craving for substance and is associated with drug use (Bandura, 1999).

Bandura (1999) described human behavior as self-regulatory in nature. Bandura stated that:

> perceived self-efficacy is the foundation of human agency. Unless people believe they can produce desired effects by their actions, they have little incentive to act. Self-efficacy beliefs promote desired changes through cognitive, motivational, affective, and choice processes. Perceived self-efficacy exerts its effects on every phase of personal change the initiation of efforts to overcome substance abuse, achievement of desired changes, recovery from relapses, and long-term maintenance of a drug-free life (Bandura, 1999, p. 240).

Perceived self-efficacy can be defined as person's beliefs about their capabilities to accomplish the designated levels of performance. Self-efficacy beliefs determine how people feel, think, motivate themselves and behave (Bandura, 1986). Self-efficacy can influence individual behavior, and then affect behavioral outcomes (Bandura, 1977). A sense of efficacy enhances human accomplishment (Bandura, 1986).

Individuals who quit smoking successfully by themselves have stronger self-efficacy than those who do not choose to quit and those who relapse. In addition, individuals who have high self-efficacy for smoking cessation can get better benefit from treatment (Bandura, 1999; Fiore et al., 2008).

Bandura (1999) provided some guidelines for treating drug abuse. He argued that perceived efficacy can help identify the individual's vulnerability and can provide guidelines for treatment (Bandura, 1999). So, for individuals who relapse, education must be provided to avoid situations that could enhance relapse. Appropriate strategies should be provided to overcome these situations, enhance the individual's self-regulatory capabilities, and teach them how to recover from setbacks (Bandura, 1999; Fiore et al., 2008). In addition, self-efficacy monitoring can evaluate the impact of substance abuse interventions (Bandura, 1999).

There are several strategies for counseling actions that health care providers can use to enhance an individual's self-efficacy for changing their behavior (Elder, Ayala and Harris. 1999). These strategies include:

> 1) increase self-efficacy for the behavior; 2) provide opportunities for the patient to master the necessary skills; 3) provide models of the targeted behavior; 4) ask the patient to rehearse the behavior and provide feedback on his/her performance; 5) address previously failed attempts and explore individual and environmental factors that may have contributed to these unsuccessful attempts; 6) explore successes with other health behavior changes and techniques employed that may generalize to the targeted behavior change; 7) increase outcome expectancies for the behavior; 8) provide information to the patient on the efficacy of the behavior; and 9) arrange for the patient to meet a similar other who has experience with the behavior and endorses its effectiveness (Elder, et al., 1999, p 277).

There are also strategies that health care providers can use to support smokers' self-efficacy for quitting smoking (Fiore et al., 2008). These strategies include: helping smokers to identify their abilities and to build on past successes in their trial to quit smoking, and offering applicable options for achievable steps for changing their smoking habits such as (telling the tobacco users to ask their health care provider or call the quit-line for advice and information, read about smoking cessation strategies, be aware of the health benefits of smoking cesstaion , initiate change in their smoking patterns, share their beliefs and ideas about smoking cessation strategies) (Fiore et al., 2008).

TREATING TOBACCO USE AND NICOTINE DEPENDENCE FRAMEWORK

Tobacco addiction is a complex phenomenon. To treat tobacco use, it is important to understand the neurobiology of nicotine dependence. Nicotine dependence results from chronic exposure to nicotine (Benowitz, 2008a; Hatsukami, et al., 2008; USDHHS, 1988, 2014). Nicotine is a major component in all tobacco forms. The pharmacologic action of nicotine is the main cause of tobacco addiction. Nicotine effects on nicotinic receptors and neurotransmitters in the brain lead to the development of neuroadaptation, tolerance to nicotine and difficulty in overcoming nicotine withdrawal symptoms in the absence of nicotine. In addition to these physiological changes, environmental conditioning is responsible for the occurrence of tobacco addiction (Benowitz, 1999, 2008a; USDHHS, 1988, 2014).

There are risk factors for starting and continuing smoking, and addiction to nicotine. Generally, these risk factors are divided into internal or external risk factors. Internal factors or characteristics of the individual experience include genes, co-morbid psychiatric disorders, and personality features. External or environmental risk factors include access to and availability of tobacco products, bans on smoking, cost of and taxes on tobacco products and family and social acceptability of smoking (Hatsukami, et al., 2008). Haberstick, et al. (2007) found that multiple genetic and individual-specific environments are important etiological factors in nicotine dependence. Susceptibility to nicotine addiction is high among persons with psychiatric disorders such as schizophrenia, alcoholism, and drug dependence (Hatsukami et al., 2008). Environmental influences, such as family members and peers smoking and advertisements or other similar media, increase the risk for smoking initiation and continuation (Broms, 2008; Hatsukami et al., 2008). Smoking is relatively common among those with a low level of education and in low socio-economic groups (Broms, 2008).

The treatment of nicotine dependence is divided into pharmacological and non-pharmacological therapy. Pharmacotherapies for nicotine dependence enhance smokers' ability to quit smoking. All smokers should be considered for pharmacotherapy, if there is no contraindication (Benowitz, 2008a; Hatsukami, et al., 2008; Fiore, et al., 2008; USDHHS, 2014). Non-pharmacological therapy- includes counseling, behavioral-educational therapy, health advice, and family support- is effective for smoking cessation (Bandura, 1999; Benowitz, 2008a; Fiore, et al., 2008; Hatsukami, et al., 2008; USDHHS,

In: Nicotine Dependence
Editor: Carlton L. Owens
ISBN: 978-1-63482-642-6

Chapter 5

SECONDHAND SMOKE EXPOSURE IN CHILDREN WITH CHRONIC LUNG DISEASE

Wei Wang*[1], *Aleksandra Bacewicz, MPH*[2], *Judy Ashouri*[1], *Randa K. Elmallah, MD, and Mai K. ElMallah**, *MD

[1]College of Medicine, University of Florida, Gainesville, FL, US
[2]University of South Florida Morsani College of Medicine, Tampa FL, US
[3]Rubin Institute for Advanced Orthopedics, Sinai Hospital of Baltimore
[4] Department of Pediatrics, College of Medicine,
University of Massachusetts, US
Funding Source: Parker B Francis Fellowship
and 1 K08 HD077040-01A1 (MKE)

ABSTRACT

Tobacco smoke is a commonly encountered air pollutant linked to chronic respiratory diseases, cardiovascular disease, and cancer in non-smokers. Non-smokers are exposed to tobacco smoke through secondhand smoke (SHS) exposure. Children are the most vulnerable

* Corresponding Author: Mai ElMallah, MD, Department of Pediatrics, College of Medicine, University of Massachusetts, 55 Lake Avenue North, Worcester, MA 01655. E-mail: mai.elmallah@umassmemorial.org.

population who are afflicted by SHS. Nationally, more than 18% of 3 to 11 year olds live with a smoker. More than 50% of children in the U.S. are exposed to second hand smoke. Children between the ages of 3 to 11 years have the highest levels of SHS exposure (60.5%), likely because they spend a majority of their time in close proximity to a smoking caregiver. In epidemiological studies, SHS exposure decreases lung growth and increases the incidence of sudden infant death, respiratory illness, middle-ear disease, wheezing, and asthma. Children with chronic lung disease are a particularly vulnerable population who are significantly affected by SHS. For example, children with asthma and cystic fibrosis are negatively impacted by SHS. Asthma is the most common childhood chronic illness, affecting 9.6% of youth in the U.S. SHS exposure in children with asthma is related to increased frequency and severity of asthma and asthma exacerbations. Similarly, children with cystic fibrosis who are exposed to SHS have an increased number of pulmonary exacerbation and hospitalizations. This chapter will address the impact of SHS on children with chronic lung disease, and interventions health care workers can implement to decrease SHS exposure in these children.

INTRODUCTION

Cigarette smoke contains greater than 4000 compounds that include many human carcinogens and toxic substances. Several of these compounds contribute to a high concentration of free radicals and oxidizing agents (Sopori 2002, Sun, Ito et al., 2014). Tobacco smoke inhaled first hand then released into the environment is known as mainstream smoke, whereas sidestream smoke is that which is generated during the burning of the cigarette during and between inhalations. During sidestream smoke, the concentration of carcinogens and toxic compounds is increased and exceeds that of mainstream smoke. Secondhand smoke (SHS) includes exhaled mainstream smoke and sidestream smoke, and is synonomous with passive or passive smoking, where a non-smoker is involuntarily inhaling products from a nearby smoker. Young children with chronic lung disease are at high risk for SHS exposure within their homes, because up to 90% of their time spent at home. This chapter will focus on the impact of exposure to SHS in children with a focus on the impact this has on the child with chronic lung disease.

Lung Physiology and SHS

The respiratory system depends on an intricate coordination of normal ciliary movement, ion transport and mucosal secretions. The airway is the first line of defense against inhaled pollutants and pathogens from the environment, and plays a vital role in innate immunity. A major component of this innate immune protection is mucociliary clearance (Liu and Di 2012). Mucociliary clearance is largely dependent on adequate mucus production and integrity of the cilia. The proximal portion of the upper (or conducting airway) through the bronchi is lined with pseudostratified columnar epithelium, subdivided into the three cell types: basal cells, ciliated cells and non-ciliated secretory cells (Ganesan, Comstock et al., 2013). The epithelium changes to simple cuboidal as the airway progresses from the conducting airways to the respiratory bronchioles, with the gradual decrease of ciliated, goblet and basal cells and the appearance of Clara cells Clara cells regulate potentially harmful proteinases secreted by neutrophils during inflammation which in turn prevents excessive tissue damage. In addition, they produce bronchiolar surfactant and antiproteases in order to regulate bronchiolar epithelial integrity and immunity, and when exposed to cigarette smoke, they metabolize aromatic hydrocarbons. In the more proximal airways, cilia and mucus have a vital role in clearing inhaled foreign elements.

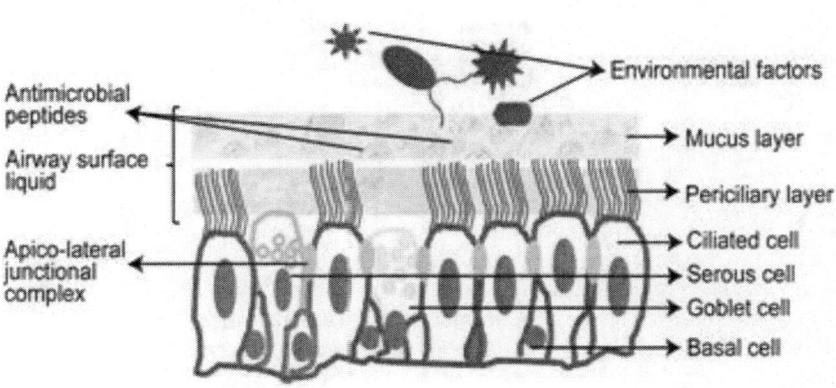

Figure 1. Airway tract epithelium: cilia and airway surface fluid trap and extrude inhaled irritants by way of mucociliary clearance. Apicolateral junctional complexes regulate permeability to prevent paracellular further invasion of irritants. Antimicrobial peptides present in the mucus layer kill inhaled pathogens. (Ganesan 2013: open access (Ganesan, Comstock et al., 2013)).

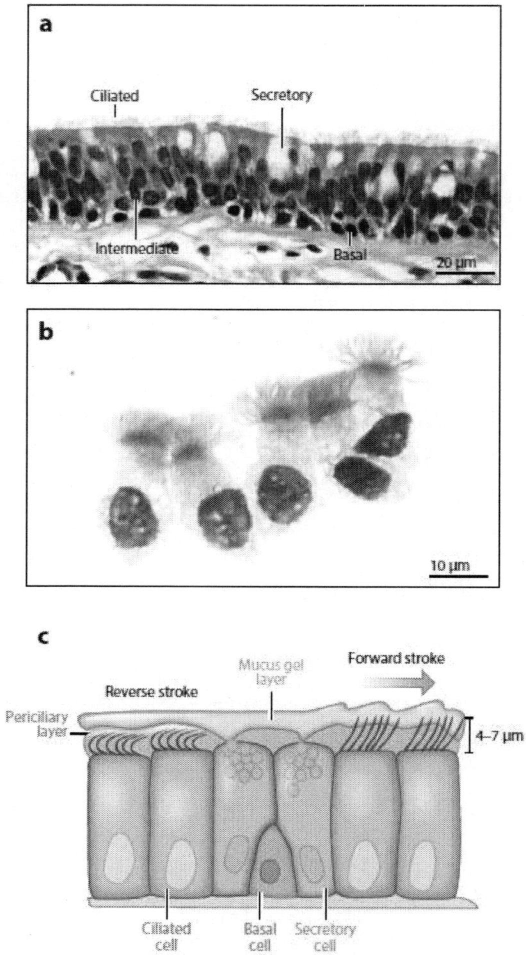

Figure 2. The human airway epithelium: (a) Histology of the large airway epithelium from a healthy nonsmoker. Ciliated cells, secretory cells, intermediate undifferentiated cells, and basal cells are visible. (Cells visualized by hematoxylin and eosin.) (b) Ciliated cells isolated from the human small airway epitheliumfrom a healthy nonsmoker. (Cells visualized by Diff-Quik.) (c) The mucociliary escalator is comprised of the mucus gel layer, the periciliary layer, and ciliated cells. The mucins produced by mucous secretory (goblet) cells are major components of the mucus gel layer, which traps microorganisms and inhaled irritants to transport them out of the lung by mucociliary clearance The distribution of water between the mucus and periciliary layers is modulated by mucins. During the power forward stroke, the tops of the cilia reach into the mucus gel layer topropel the mucus upward and out of the lungs. During the slow return stroke, the cilia recede and are contained within the periciliary layer. Normal cilia length (4–7 μm, depending on the airway region) is critical for effective mucociliary clearance. (Tilley 2015 (Tilley, Walters et al., 2015)).

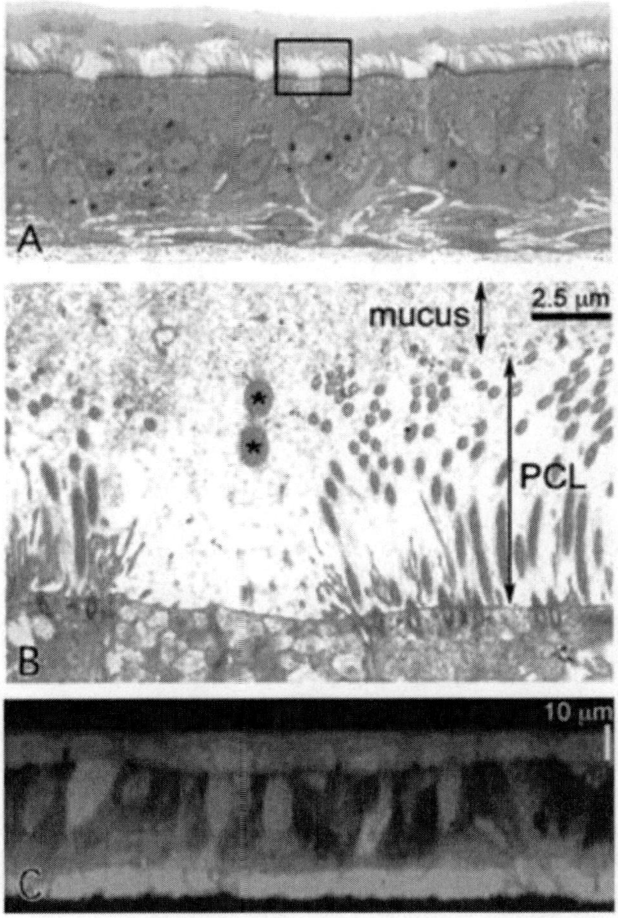

Figure 3. A periciliary liquid layer (PCL) is essential for mucus transport in the airways. (A) This layer is shown as a clear layer between air and liquid in a normal culture of human tracheobronchial epithelial cells. (Epon-embedding, Richardson's stain.) (B) TEM of the boxed in portion in (A) There is a layer of dense mucus above the PCL. (The asterisks denote two polystyrene beads added to track the movement of mucus.) (C) The PCL as viewed in a live cell preparation by X-Z plane confocal microscopy. Airway surface liquid (ASL) is labeled with Texas red dextran on the apical surface (red) and the cells were labeled with calcein (green). (Livraghi and Randell 2007)).

Cilia are a vital component of mucociliary clearance. The ciliated cells of the respiratory system line the majority of the conducting aiways and are critical for the clearance of unwanted materials from the lungs. Approximately

250 to 300 cilia are found on a ciliated cell apical surface, with an average length of 6 micrometers in the upper airway, decreasing in length to about 3.5 micrometers in the terminal bronchioles. The cilia have an extensive mitochondrial supply immediately below the apical epithelial surface that provides energy for their coordinated ciliary movement. This powerful combination, the fluid mucosal layer and functioning cilia, allows for a mechanism in which the mucus traps foreign particles, and the cilia beat the mucus upward in wave-like fashion to ultimately push unwanted material into the pharynx for swallowing or expulsion via coughing.

Mucus production is an integral part of our defense mechanism and is vital for a functionally normal mucociliary clearance. Mucus is secreted from the non-ciliated secretory cells of the airway, which include the goblet cells of the surface epithelia and mucous cells of submucosal glands and it is comprised of water and ions, mucin glycoproteins and macromolecules that serve antimicrobial and antiprotease functions (Adler, Tuvim et al., 2013). Mucus is an important component of the airway surface liquid which is composed of mucus and a periciliary liquid layer. The periciliary liquid layer is a less viscous layer of fluid in between the cilia and the mucosal layer. Hydration of the liquid layers and the interlinking of mucin glycoproteins create a physical barrier for pathogens and particulate matter to get trapped, thereby preventing the further penetration of foreign objects. A more hydrated periciliary liquid layer results in more efficient mucociliary clearance of foreign particles.

Mucociliary Clearance Dysfunction

Cigarette smoke interferes with the delicate balance of mucus secretion and ciliary function. Mucociliary clearance is affected in one of two primary ways: 1) ciliary function is impaired or 2) the mucus is dehydrated, thereby thickened (Munkholm and Mortensen 2014). Cigarette smoke exposure, both first and second hand, induces functional and structural changes within the airway. Cilia length is reduced by 15% in the airway epithelium of healthy smokers when compared to non-smokers (Leopold, O'Mahony et al., 2009). The amount by which cilia length was decreased varied depending on the analytic method used, and ranged from a 7% to 15% decrease as compared to the non-smokers. Using RNA extraction and microarray analysis, six genes that are involved in either the structure or function of cilia within the large airway epithelium were significantly downregulated (p values ranged from <0.001 to <0.005) in the smoking group. It is estimated that only the distal

10% of cilia actually reaches the mucosal layer to facilitate mucociliary clearance, as most of the cilia length is found within the periciliary layer. With direct cigarette smoke exposure, the decreased lengths hinders the cilias' ability to adequately propel mucus upward, and this results in inadequate pathogen and toxin clearance out of the airway.

Furthermore, cigarette smoke causes more widespread damage to the mucosa (Tilley, Walters et al., 2015). In one study, nasal mucosa of children aged 5 to 11 years was biopsied and compared to a control group with negative household smoking history (Elwany, Ibrahim et al., 2012). Ultrastructural changes observed in the exposed group were as follows: epithelial changes (45%), hyperplasia of goblet cells (30%), hyperplasia of seromucinous glands (40%) and vascular congestion (45%). None of the ultrastructural changes were observed in the unexposed group. The epithelial changes included regions of patchy cilia loss, with hyperplasia and distension of goblet cells and seromucinous glands. The glands were distended and showed evidence of increased secretion with an abundance of mitochondria and endoplasmic reticulum with swollen cisternae were present. The patches of lost cilia and increased mucosal secretions would contribute to an overall decrease in mucociliary clearance of the airways. Thus, mucociliary clearance is impaired by cigarette smoke exposure, secondary to dysfunctional or reduced cilia in conjunction with mucosal abnormalities (Habesoglu, Kule et al., 2015).

Although cilia beat frequency is changed with smoke exposure, there is conflicting data about the exact impact on frequency; i.e. does it increase or decrease? (Cohen, Zhang et al., 2009, Navarrette, Sisson et al., 2012, Tilley, Walters et al., 2015). Some of these differences may be due to the various ways in which cilia beat frequency is studied and the mammals within which they are studied (Houtmeyers, Gosselink et al., 1999). *In vitro* models will differ from *in vivo* models, as *in vivo* models will have the added effects of blood flow and innervation and *in vitro* models may be affected by the experimental variables, like temperature and humidity. Baseline ciliary beat frequency obtained through bronchoscopies was determined to range from 12-15 Hz (Helleday, Huberman et al., 1995). One study examined ciliary beat frequency in the respiratory airways of groups of smokers, non-smokers, and those exposed to environmental tobacco smoke either domestically or occupationally (Zhou, Wang et al., 2009). It was found that ciliary beat frequency was increased by an average of 10-17% for those who were exposed to direct smoke and SHS. There was no detectable difference in the ciliary beat frequency between the non-smoking/environmental exposure group and the smoker group, indicating that even environmental exposure to smoke can cause significant changes in cilia function. There have been other studies examining the link between smoke exposure and ciliary beat frequency that have found the opposite results. A study examined *in vitro* human sinonasal epithelial (HSNE) and murine nasal septal epithelial (MNSE) cultures and the effects of cigarette smoke condensate on these cultures, which are comparable to those *in vivo* in human upper airways (Cohen, Zhang et al., 2009). In these cell cultures, the cigarette smoke condensate ultimately dampened the stimulation of ciliary beat frequency, which occurs via a cyclic adenosine monophosphate (cAMP)-activated pathway. These results support the hypothesis that cigarette smoke contributes to inefficient mucociliary clearance.

The comprehensive effect of cigarette smoke on mucociliary clearance is conflicting, it is either increased or decreased in differing analyses mimicking the effects of cigarette smoke (Houtmeyers, Gosselink et al., 1999, Zhou, Wang et al., 2009, Habesoglu, Kule et al., 2015). According to Zhou et al., smokers who are asymptomatic for respiratory symptoms generally report more watery mucus and increases in mucociliary clearance (Zhou, Wang et al., 2009). All participants had no respiratory symptoms, were in good overall health and none were diagnosed with a chronic respiratory disease (Zhou, Wang et al., 2009). Thus, it appears that these individuals had an increased ciliary beat frequency in response to irritants, which consequently leads to an increase in mucociliary clearance. On the other hand, individuals who are chronic smokers report a decrease in mucociliary clearance, as compared to healthy individuals (Hogg 2004). This indicates that those who have had more or longer exposure to cigarette smoke ultimately have decreased ciliary beat frequency, due to the greater exposure to cigarette smoke toxins. This spectrum of possibilities has yet to be fully researched.

Inflammation and Loss of Epithelial Integrity

Aside from proper structure and function of the components that aid in mucociliary clearance, the integrity of our airway defenses also depend on epithelial barrier integrity, cell-cell junctions, immune cells and their regulation (Martin and Frevert 2005, Xiao, Puddicombe et al., 2011). These factors of immunity work interdependently with the components of mucociliary clearance to maintain a healthy pulmonary system. The three broad classes of macrophages within the respiratory system are divided by where they are generally present: in the bronchial airways, the alveoli or the

create an inflammatory environment. Leukotrienes C4, D4, and E4 are, as a group, referred to as the cysteinyl leukotrienes. These leukotrienes bind to the receptors CysLT1 and CysLT2. CysLT1 is involved in smooth muscle contraction in the airways, chemotaxis, and increasing vascular permeability. CystLT2 is involved in constriction of smooth muscle in the pulmonary vessels. These cysteinyl leukotrienes are activated during asthma attacks. The effects of LTE4 are better understood due to the effectiveness of LTE4 inhibitors in helping improve the symptoms of asthma. Montelukast and Zafirlukast are a CystLT1 receptor antagonists, which act to limit airway hyperresponsiveness (Drazen, Israel et al., 1999). Studies have shown Montelukast to be effective in decreasing asthma symptoms in children who are smoked to SHS smoke (Lazarus, Chinchilli et al., 2007, Rabinovitch, Strand et al., 2008). It also decreased the number of days with asthma symptoms and the need for other asthma medications such as beta-agonists and oral steroids. These results have been demonstrated in children as young at 2 years old (Knorr, Franchi et al., 2001). Thus, elevated LTE4 levels in the urine are indicative of poor asthma control.

Furthermore, there appears to be an association between increased uLTE4 levels and a greater need for albuterol inhalers (Rabinovitch, Strand et al., 2008). Moreover, these children are at greater risk of asthma exacerbations even with the use of inhaled corticosteroids. Interestingly, while tobacco exposure was associated with increased exacerbation risk, factors such as nighttime symptom frequency, lung function pre- and post-bronchidilator, and exhaled nitric oxide levels did not show the same association with exacerbations as did $uLTE_4$ levels (Rabinovitch et al, 2011).

Social, Environmental, and Genetic Factors

In the United States, greater than 30% children are exposed to SHS in and around their homes (Butz, Breysse et al., 2011). Unfortunately, this number is higher in young inner city children with asthma with over 50% of these children being exposed to SHS in their homes (Delva, Tellez et al., 2005). This impacts young preschool children more than any other age group because they spend up to 90% of their time at home (Klepeis, Nelson et al., 2001). Often, the caregiver of the child is the main smoker which results in much greater SHS exposure as a result of close proximity and prolonged exposure (Winickoff, Berkowitz et al., 2005). SHS has been associated with increased risk for the development and severity of asthma and difficulty in managing asthma symptoms. Factors that are associated with increased exposure to SHS included caregivers with increased stress and/or depressive symptoms (Butz, Halterman et al., 2011). Daily stress and asthma stress was examined using a questionnaire which had caregivers select a number between 0 to 10 to indicated their level of routing daily stress. Daily stress was high among inner city families and correlated with increased asthma prevalence in these homes. However, asthma stress was not high presumably because the daily financial stresses were more than the stress of managing a child's asthma (Butz, Halterman et al., 2011). This highlights the importance of addressing psychosocial support in smoking cessation counseling of families with a lower socioeconomic status. In parallel, caregivers were given a depression questionnaire to gauge their level of depression and this was compared with their child's asthma symptoms. Furthermore, 3-5 year olds had a higher salivary cotinine concentration, a measure of tobacco exposure, than older children (Butz, Breysse et al., 2011). Of note, young children still had high salivary cotinine levels even if there was a household ban on smoking (i.e. no smoking inside the home), most likely because the young child could not be left alone and had to accompany the caregiver who was smoking outside the home (Butz, Halterman et al., 2011).

Effects of Maternal, Paternal and Multiple Smokers

The primary household smoker is often the primary caregiver of the child with asthma and is usually also the mother in the home (Butz, Breysse et al., 2011). SHS exposure to maternal smoking is associated with childhood asthma, as well as increased wheeze, cough, chronic phlegm, and breathlessness (Cook and Strachan 1997, Winickoff, Berkowitz et al., 2005). The adverse effects of parental smoking on children's respiratory symptoms have been shown in both preschool and school aged children. Maternal smoking exposure in the postnatal period appears to have a larger effect than later exposure, which emphasizes the need for mothers to quit smoking as early as possible, ideally before pregnancy (Wall 2000, Neuman, Hohmann et al., 2012). In fact, maternal smoking was associated with a 52% increased risk of wheeze and 24% increased risk of asthma (Burke, Leonardi-Bee et al., 2012). The greater negative effects of smoking exposure during infancy and preschool may be due to increased exposure as these children spend more time with their mothers than older school aged children, increased maturity of the respiratory system, or even a combination of the two (Cook and Strachan 1997, den Dekker, Sonnenschein-van der Voort et al., 2015).

In addition, there is a dose-response relationship between smoking and wheezing, as maternal smoking of more than 10 cigarettes per day is associated with a greater incidence of wheezing before 3 years old and continued wheeze through 6 years old (Neuman, Hohmann et al., 2012). There also exists data suggesting an increased incidence of asthma by the age of 5 years in children whose mothers smoke (Strachan and Cook 1997, Kopp, Sarzynski et al., 2015). The effect is seen in children without diagnosed asthma as well: wheezing syndrome in children between 6 and 23 months of age is associated with exposure to SHS. Again, this association increases as the number of cigarettes smoked increases, as measured by maternal reported data and children's urine cotinine level (Schvartsman, Farhat et al., 2013). Studies suggest that paternal smoking, similar to maternal smoking, is associated with increased prevalence of asthma, cough, and wheeze. While not as marked as the effects of maternal smoking, the negative effects of paternal smoking are still significant (Cook and Strachan 1997, Burke, Leonardi-Bee et al., 2012). A dose response between parental smoking, childhood asthma and respiratory symptoms exists: as the number of smokers in the home increases so does prevalence of asthma and its symptoms such as wheeze, cough, chronic phlegm, and breathlessness (Cook and Strachan 1997). Notably, as the number of household smokers increased, so did the salivary cotinine concentrations of children with asthma (Butz, Halterman et al., 2011), providing biochemical evidence of greater exposure.

SHS Exposure in the Child with Cystic Fibrosis

Cystic fibrosis (CF) is a common autosomal recessive disease that affects at least 30,000 people in the US with an approximate incidence of 1 in 3,400 births. CF is caused by a mutation in the cystic fibrosis transmembrane receptor (CFTR) gene. CFTR is a chloride channel and defects in the CFTR gene result in abnormal transport of chloride and sodium in many different organs including the lungs and the pancreas. A consequence of this abnormal chloride and sodium flow in the lungs is defective water transportation across cells that line the respiratory tract; thus resulting in a reduction in the fluid lining the airways. The decreased fluid causes thickened secretions in the airways leading to airway obstruction and frequent respiratory infections. As a result, lung disease is one of the leading cause of morbidity and mortality in

CF patients (Strausbaugh and Davis 2007). Furthermore, aberrant CFTR expression in the gastrointestinal tract and pancreas impairs the ability of pancreatic enzyme secretion, causing malabsorption. This leads to malnourishment and vitamin deficiency in children with cystic fibrosis.

SHS exposure is associated with lung disease and malnutrition (Strachan and Cook 1997). In the setting of CF the effects of SHS exposure are compounded and are critical to address. The following section will summarize the prevalence of SHS in CF, the intricate interaction between genetic and environmental factors influenced by SHS, and the impacts of SHS on pulmonary disease and growth parameters in CF patients.

Prevalence

Smoke exposure is common in the CF population among caregivers and patients afflicted with CF. Nationally, in 2012, the CF registry reported a 20% exposure rate of smoking in children under 18 years of age (Cystic Fibrosis Foundation 2013). Unfortunately, the prevalence of smoke exposure in CF children is currently on the rise after there had been a steady downward trend from 2006 to 2011 (see figure 1) (Cystic Fibrosis Foundation 2013). These smoke exposure data in the CF patient registry are dependent on self-reported accounts by caregivers and not objective measures, and thus, they are most likely under stated. According to Kohler et al, objective measures such as urine cotinine levels revealed a 65% higher level of SHS exposure in children with cystic fibrosis than that reported by caregivers (Kohler, Sollich et al., 1999, Cystic Fibrosis Foundation 2013). This implies that the CF registry results are most likely lower than the true prevalence. The discrepancy between parental reporting and objective measures was only 18% when their children did not have CF or any other chronic lung disease. This may be due to the tendency of parents to understate or withhold the truth about SHS exposure when questioned by the physician or other healthcare professionals in the context of an illness of their children, perhaps in fear of embarrassment. The degree of parental under reporting of SHS when a child has CF is worse when the burden of guilt is higher (Gilljam, Stenlund et al., 1990, Verma, Clough et al., 2001). One study showed that 80% of parents had made an attempt at smoking cessation when their child was diagnosed with CF [15]. Therefore, many parents of CF patients are well aware of the detrimental effects of SHS exposure, thus, it is understandable for parents to underreport their smoking behavior when faced with the guilt of continuing to smoke in the face of their child's illness and for some, the reminder of their failed cessation attempt when questioned about smoking behavior. Nonetheless, several studies assessing the percentage of SHS exposure in CF patients have demonstrated that the parental reporting rate varies between 23-56%(Rubin 1990, Collaco, Vanscoy et al., 2008), and that SHS may actually be higher in CF children than in children without chronic lung disease (Smyth, O'Hea et al., 1994, Verma, Clough et al., 2001).

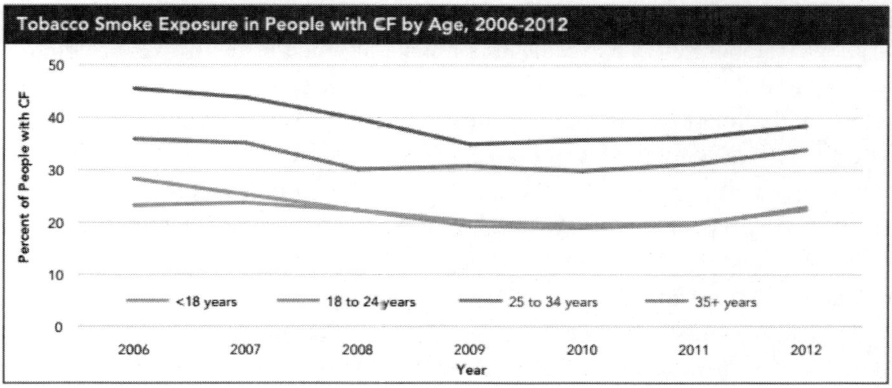

Figure 5. A downward trend in tobacco exposure from 2006-2011 is demonstrated , however, an upward rise in tobacco smoke exposure is observed from 2011-2012. (Cystic Fibrosis Foundation 2012).

In addition to SHS, in utero exposure to smoking is a problem in newborns with CF. Approximately 16.5-44% of parents of CF patients actively smoked during the pregnancy. Thus, it is clear that SHS is highly prevalent in this vulnerable population of patients, and that most of the exposure is from parents and caregivers of CF patients (Rubin 1990, Smyth, O'Hea et al., 1994, Verma, Clough et al., 2001, Collaco, Vanscoy et al., 2008, Kopp, Sarzynski et al., 2015). Kopp et al., surveyed caretakers of 75 CF infants and the results revealed that most (73%) of SHS exposed infants were exposed by either their mother (33%) or both parents (39.4%).

Impact of Smoke Exposure on CFTR

In CF, changes associated with mucociliary clearance are of particular importance (see mucociliary clearance dysfunction above). The CFTR gene encodes for a cyclic adenosine monophosphate (cAMP)-activated anion transmembrane channel found in the apical membrane cells of the surface airway epithelium and submucosal glands. The channel mainly transports Cl^- ions, in addition to regulating movement of other ions, like HCO_3^-, in the epithelium. There are over 1400 recognized CFTR mutations, with the most common being a deletion of three base pairs, resulting in the loss of a phenylalanine in the 508 position—a deletion known as ΔF508. The mutations range from mild to severe, as some mild mutations can cause dysfunction in other organs systems without any signs, or few, of respiratory disease. The most severe mutations may result in complete absence of the CFTR protein at the apical surface of the epithelium. Although CF generally affects multiple organs, its hallmark is the inability of the CFTR protein to properly regulate the transport of ions and water across epithelial surfaces. A result of this abnormality is the presence of thickened, dehydrated mucosal and periciliary layer, which tends to lead to obstruction of tubular structures and ducts within the body (Livraghi 2007).. Additionally, individuals with CF tend to have respiratory dysfunction early in life, as their lungs are easily colonized with bacteria, particularly *Haemophilus influenza, Staphylococcus aureus and Pseudomonas aeruginosa.* Infection leads to persistent inflammation, adding to the present obstruction of airways by dehydrated airway surface liquid (Berkebile 2014).

Since smoke exposure causes negative consequences on both cilia function and adequate makeup of the airway surface fluid, the presence of CF in conjunction with smoke exposure likely leads to worse health outcomes. A study examined human bronchial epithelial cells (HBECs) and the acute effect of cigarette smoke on the airway surface liquid (Clunes, Davies et al., 2012). The HBECs were exposed to cigarette smoke from one cigarette and compared to controls. There was a reduction in the volume and height of the airway surface fluid in the exposed cells, which lasted over 2.5 hours. The mechanism for this was elucidated to be a diminished CFTR function, as this anion channel was internalized upon smoke exposure. Specifically, cigarette smoke induces cytosolic translocation of calcium ions, which signals internalization of CFTR receptors from the plasma membrane. In addition, with chronic exposure, cytosolic calcium levels steadily raised with no desensitization, implying a dose-dependent effect on worsening CFTR dysfunction (Rasmussen, Sheridan et al., 2014).

These results are in line with other studies that show a slower time of mucociliary clearance in individuals with CF, likely due to the inadequate hydration of the mucus and the periciliary liquid layer, normally regulated by

CFTR (Ratjen 2006). Persistent infections and difficulty expelling bacteria are prominent in CF and made worse by SHS smoke. In addition, there was a lower rate of bacterial clearance in a study examining mice infected with *Pseudomonas aeruginosa* after cigarette smoke exposure as compared to murine controls. This depressed rate was found to have an association with an increase in inflammation throughout the airway, with increased numbers of chemokines and cytokines (Drannik, Pouladi et al., 2004).

In addition, CFTR is suppressed by cigarette smoke exposure both in vitro and in vivo (Verma, Clough et al., 2001). Cantin et al., measured CFTR expression in two cell lines exposed to cigarette smoke (Cantin, Hanrahan et al., 2006). Cigarette smoke decreased CFTR mRNA levels ($p<0.0001$), protein levels ($p<0.01$) and CFTR expression ($p<0.001$) in vitro. A normal nasal transepithelial difference is an indicator an intact CFTR mediated mechanism of sodium transport at the apical surface of respiratory epithelial cells. The same study by Cantin et al., revealed that the CFTR-mediated nasal transepithelial difference is -9.6 ± 4.0 in smokers and -22.3 ± 10.1 in nonsmokers, showing a significant decrease in the smoking group ($p<0.001$) (Cantin, Hanrahan et al., 2006). These data also suggest that cigarette smoke may induce an acquired CFTR defect in non-CF patients which may be responsible for the respiratory clinical manifestations of smokers, and could potentially lead to further CFTR dysfunction in CF patients.

Another recent study by Raju et al., investigated whether cigarette smoking impacted the extrapulmonary manifestations of the CFTR function (Raju, Jackson et al., 2013). Sweat chloride levels were elevated in healthy smokers, former smokers, and smokers with COPD as compared with normal subjects.

In addition, CFTR function assessed through intestinal current measurements yielded a 65% decrease in smokers as compared to controls. The same study also revealed a similar decrease in CFTR function in bronchial epithelial cells, as determined by current measurements in isolated electrical chambers, after exposure to smokers' plasma (Raju, Jackson et al., 2013). The decreased CFTR function in extra-pulmonary organs such as the sweat glands and intestinal epithelium as observed in this study may explain the similarity in non-respiratory clinical features (malnutrition, pancreatitis, infertility) between smokers and CF patients.

Genetics of CF and Smoke Exposure

Several other studies have attempted to demonstrate an association between SHS exposure and severity of clinical disease in different genetic populations of CF patients. Smoke exposure is associated with a significant percentile decrease in pulmonary lung function (predicted $FEV_{1)}$ in *CFTR* non-ΔF508 homozygotes, *TGFβ1*−509 TT homozygotes, and *TGFβ1* codon 10 CC homozygotes (Collaco, Vanscoy et al., 2008). Individuals carrying mutations other than ΔF508 or just one copy of the ΔF508 gene had a significant reduction of pulmonary function by 13 percentile points when exposed to SHS smoke. More drastically, individuals carrying 2 copies of the T allele (TT homozygotes) at the -509 promoter site of *TGFβ1* had a reduction of lung function by 22 percentile points as compared to TT homozygotes not exposed to SHS smoke. Similarly, in individuals with CC homozygotes on codon 10 on *TGFβ1; lung function declined by 20 percentile points as compared to those who were not exposed. When patients with different CF genotypes were pooled together, the pulmonary function scores of the exposure group as a whole declined by 9.8 percentile points compared to those not exposed. (Collaco, Vanscoy* et al., *2008).*

Another study demonstrated significantly poorer lung function and hospitalizations in ΔF508 homozygotes exposed to SHS (Campbell, Parker et al., 1992). This study only examined 44 patients homozygous for the ΔF508 mutation. Results of this study showed a significance towards more pulmonary-related hospitalizations and lower pulmonary function scores in participants exposed to SHS. However, two better powered studies failed to replicate these findings in ΔF508 homozygotes (Kovesi, Corey et al., 1993, Collaco, Vanscoy et al., 2008). The Collaco study performed a linear regression analysis which revealed that F508 homozygotes' lung function scores were not reduced when exposed to SHS smoke ($p=0.11$) (Collaco, Vanscoy et al., 2008). Similarly, Kovesi et al., detected no significant difference in FEV_1 or weight percentile in their subgroup of ΔF508 homozygotes with SHS smoke exposure (Kovesi, Corey et al., 1993). Based on the current findings, it seems like there is no significant association between lung function or clinical status when ΔF508 homozygotes are exposed to SHS smoke. Since there is adequate evidence that CFTR regulates the detrimental effect of smoke exposure in CF and non-CF patients (see above section), it is logical that the subgroup that lacks any residual CFTR function (such as ΔF508 homozygotes) are therefore missing a main target

levels can be used as a screening tool for SHS exposure in this vulnerable population, whether in the hospital or clinic, as it is a rapid and non-invasive test that is easy to interpret. This information can possibly help healthcare providers tailor the clinic visit with more emphasis on discussion of smoking cessation.

Smoking cessation counseling specific to the CF population were briefly addressed in two other studies. Data gathered through questionnaires administered to patients in a Manchester CF unit revealed that the vast majority of parents, spouse, siblings and friends of CF patients who smoked wanted smoking cessation advice and support to help them stop (Verma, Clough et al., 2001). The authors of this study also recommended screening for active and SHS exposure with urinary cotinine, since a vast 76% of their patients in the unit were exposed to SHS, with 4 of which who reported being exposed in the same room while performing their breathing treatments at home. This study did not address the impact of smoking cessation nor did it address whether or not these caretakers received smoking cessation advice despite the large prevalence of SHS smoke exposure.

A small study conducted at a tertiary clinic assessed smoking cessation rates among caretakers of patients at a tertiary pulmonary clinic. The majority of patients were asthmatics, with a small subset of CF patients. Caretakers were asked to complete a questionnaire about their smoking habits, and healthcare providers subsequently assessed the caretakers' readiness to quit and proactively enrolled them in a telephone smoking cessation program through fax. Enrolled caretakers were contacted by trained representatives of the program and given the choice to enroll in a telephone, web-based, or face-to-face smoking cessation counseling program. Caretakers of 30 CF patients completed the questionnaire, 11 of whom reported active smoking. However, only 3 out of the 11 (27%) CF caretakers were willing to be referred to the smoking cessation program, in contrast to the 19 out of 39 (48%) caretakers of asthma patients. It is quite interesting that a smaller percentage of smoking caretakers of CF patients were interested in smoking cessation, as opposed to the caretakers of asthmatic children. Perhaps these parents needed nicotine to cope with the everyday stress of taking care of a child with CF. It is also possible that some caretakers of CF children ceased smoking once they learned of their child's CF diagnosis (Bacewicz, Wang et al., 2014).

It is apparent that there is a lack of targeted efforts at smoking prevention or counseling of caretakers of CF patients, and more studies are needed to assess the effectiveness of a tailored counseling program towards this population. One study reported that over 80% of caretakers of asthmatic and

CF children reported a change in their smoking behavior after learning of their child's diagnosis, as compared to only 40% of caretakers of children with non-respiratory illnesses (Butz and Rosenstein 1992). Therefore, future counseling efforts need to focus on the benefits of smoking cessation on childrens' health.

REFERENCES

Adler, K. B., M. J. Tuvim and B. F. Dickey (2013). "Regulated mucin secretion from airway epithelial cells." *Front Endocrinol (Lausanne)* 4: 129.

Bacewicz, A., W. Wang, J. Ashouri and M. K. ElMallah (2014). "Children with Chronic Lung Disease: Facilitating Smoking Cessation for their Caregivers." *J Community Health.*

Balhara, J. and A. S. Gounni (2012). "The alveolar macrophages in asthma: a double-edged sword." *Mucosal Immunol* 5(6): 605-609.

Beydon, N., F. Amsallem, M. Bellet, M. Boule, M. Chaussain, A. Denjean, R. Matran, I. Pin, C. Alberti and C. Gaultier (2002). "Pulmonary function tests in preschool children with cystic fibrosis." *American journal of respiratory and critical care medicine* 166(8): 1099-1104.

Beydon, N., F. Amsallem, M. Bellet, M. Boule, M. Chaussain, A. Denjean, R. Matran, I. Pin, C. Alberti and C. Gaultier (2002). "Pulmonary function tests in preschool children with cystic fibrosis." *Am J Respir Crit Care Med* 166(8): 1099-1104.

Bodnar, R., L. Kadar, K. Holics, R. Ujhelyi, L. Kovacs, K. Bolbas, G. Szekely, K. Gyurkovits, E. Solyom and A. Meszaros (2014). "Factors influencing quality of life and disease severity in Hungarian children and young adults with cystic fibrosis." *Italian journal of pediatrics* 40: 50.

Botelho, F. M., G. J. Gaschler, S. Kianpour, C. C. Zavitz, N. J. Trimble, J. K. Nikota, C. M. Bauer and M. R. Stampfli (2010). "Innate immune processes are sufficient for driving cigarette smoke-induced inflammation in mice." *Am J Respir Cell Mol Biol* 42(4): 394-403.

Boyce, J. A., D. Broide, K. Matsumoto and B. S. Bochner (2009). "Advances in mechanisms of asthma, allergy, and immunology in 2008." *J Allergy Clin Immunol* 123(3): 569-574.

Burke, H., J. Leonardi-Bee, A. Hashim, H. Pine-Abata, Y. Chen, D. G. Cook, J. R. Britton and T. M. McKeever (2012). "Prenatal and passive smoke exposure and incidence of asthma and wheeze: systematic review and meta-analysis." *Pediatrics* 129(4): 735-744.

Butz, A. M., P. Breysse, C. Rand, J. Curtin-Brosnan, P. Eggleston, G. B. Diette, D. Williams, J. T. Bernert and E. C. Matsui (2011). "Household smoking behavior: effects on indoor air quality and health of urban children with asthma." *Matern Child Health J* 15(4): 460-468.

Butz, A. M., J. S. Halterman, M. Bellin, M. Tsoukleris, M. Donithan, J. Kub, R. E. Thompson, C. L. Land, J. Walker and M. E. Bollinger (2011). "Factors associated with second-hand smoke exposure in young inner-city children with asthma." *J Asthma* 48(5): 449-457.

Butz, A. M. and B. J. Rosenstein (1992). "Passive smoking among children with chronic respiratory disease." *The Journal of asthma : official journal of the Association for the Care of Asthma* 29(4): 265-272.

Campbell, P. W., 3rd, R. A. Parker, B. T. Roberts, M. R. Krishnamani and J. A. Phillips, 3rd (1992). "Association of poor clinical status and heavy exposure to tobacco smoke in patients with cystic fibrosis who are homozygous for the F508 deletion." *J Pediatr* 120(2 Pt 1): 261-264.

Campbell, P. W., 3rd, R. A. Parker, B. T. Roberts, M. R. Krishnamani and J. A. Phillips, 3rd (1992). "Association of poor clinical status and heavy exposure to tobacco smoke in patients with cystic fibrosis who are homozygous for the F508 deletion." *The Journal of pediatrics* 120(2 Pt 1): 261-264.

Cantin, A. M., J. W. Hanrahan, G. Bilodeau, L. Ellis, A. Dupuis, J. Liao, J. Zielenski and P. Durie (2006). "Cystic fibrosis transmembrane conductance regulator function is suppressed in cigarette smokers." *Am J Respir Crit Care Med* 173(10): 1139-1144.

CDC (2012). "Summary Health Statistics for U.S. Children: National Health Interview Survey."

Clunes, L. A., C. M. Davies, R. D. Coakley, A. A. Aleksandrov, A. G. Henderson, K. L. Zeman, E. N. Worthington, M. Gentzsch, S. M. Kreda, D. Cholon, W. D. Bennett, J. R. Riordan, R. C. Boucher and R. Tarran (2012). "Cigarette smoke exposure induces CFTR internalization and insolubility, leading to airway surface liquid dehydration." *FASEB J* 26(2): 533-545.

Cohen, N. A., S. Zhang, D. B. Sharp, E. Tamashiro, B. Chen, E. J. Sorscher and B. A. Woodworth (2009). "Cigarette smoke condensate inhibits transepithelial chloride transport and ciliary beat frequency." *Laryngoscope* 119(11): 2269-2274.

Collaco, J. M., L. Vanscoy, L. Bremer, K. McDougal, S. M. Blackman, A. Bowers, K. Naughton, J. Jennings, J. Ellen and G. R. Cutting (2008).

"Interactions between secondhand smoke and genes that affect cystic fibrosis lung disease." *JAMA* 299(4): 417-424.

Cook, D. G. and D. P. Strachan (1997). "Health effects of passive smoking. 3. Parental smoking and prevalence of respiratory symptoms and asthma in school age children." *Thorax* 52(12): 1081-1094.

Cystic Fibrosis Foundation (2013). Cystic Fibrosis Patient Registry Annual Data Report 2012.

Delva, J., M. Tellez, T. L. Finlayson, K. A. Gretebeck, K. Siefert, D. R. Williams and A. I. Ismail (2005). "Cigarette smoking among low-income African Americans: a serious public health problem." *Am J Prev Med* 29(3): 218-220.

den Dekker, H. T., A. M. Sonnenschein-van der Voort, J. C. de Jongste, I. K. Reiss, A. Hofman, V. W. Jaddoe and L. Duijts (2015). "Tobacco smoke exposure, airway resistance and asthma in school-age children: The Generation R Study." *Chest.*

Drannik, A. G., M. A. Pouladi, C. S. Robbins, S. I. Goncharova, S. Kianpour and M. R. Stampfli (2004). "Impact of cigarette smoke on clearance and inflammation after Pseudomonas aeruginosa infection." *Am J Respir Crit Care Med* 170(11): 1164-1171.

Drazen, J. M., E. Israel and P. M. O'Byrne (1999). "Treatment of asthma with drugs modifying the leukotriene pathway." *N Engl J Med* 340(3): 197-206.

Eisner, M. D., E. H. Yelin, P. P. Katz, G. Earnest and P. D. Blanc (2002). "Exposure to indoor combustion and adult asthma outcomes: environmental tobacco smoke, gas stoves, and woodsmoke." *Thorax* 57(11): 973-978.

Elwany, S., A. A. Ibrahim, Z. Mandour and I. Talaat (2012). "Effect of passive smoking on the ultrastructure of the nasal mucosa in children." *Laryngoscope* 122(5): 965-969.

Flouris, A. D., C. I. Vardavas, G. S. Metsios, A. M. Tsatsakis and Y. Koutedakis (2010). "Biological evidence for the acute health effects of secondhand smoke exposure." *Am J Physiol Lung Cell Mol Physiol* 298(1): L3-L12.

Ganesan, S., A. T. Comstock and U. S. Sajjan (2013). "Barrier function of airway tract epithelium." *Tissue Barriers* 1(4): e24997.

Gilljam, H., C. Stenlund, A. Ericsson-Hollsing and B. Strandvik (1990). "Passive smoking in cystic fibrosis." *Respiratory medicine* 84(4): 289-291.

Habesoglu, T. E., M. Kule, Z. G. Kule, H. S. Deveci, A. Yaylaci, A. O. Gursel and M. Habesoglu (2015) "How does parental smoking affect nasal mucociliary clearance in children?" *Eur Arch Otorhinolaryngol* 272(3): 607-611.

Heijink, I. H., S. M. Brandenburg, D. S. Postma and A. J. van Oosterhout (2012). "Cigarette smoke impairs airway epithelial barrier function and cell-cell contact recovery." *Eur Respir J* 39(2): 419-428.

Helleday, R., D. Huberman, A. Blomberg, N. Stjernberg and T. Sandstrom (1995). "Nitrogen dioxide exposure impairs the frequency of the mucociliary activity in healthy subjects." *Eur Respir J* 8(10): 1664-1668.

Hernandez-Alvidrez, E., G. Alba-Reyes, B. C. Munoz-Cedillo, J. L. Arreola-Ramirez, M. E. Furuya, M. Becerril-Angeles and M. H. Vargas (2013). "Passive smoking induces leukotriene production in children: influence of asthma." *J Asthma* 50(4): 347-353.

Hogg, J. C. (2004). "Pathophysiology of airflow limitation in chronic obstructive pulmonary disease." *Lancet* 364(9435): 709-721.

Houtmeyers, E., R. Gosselink, G. Gayan-Ramirez and M. Decramer (1999). "Regulation of mucociliary clearance in health and disease." *Eur Respir J* 13(5): 1177-1188.

Hussell, T. and T. J. Bell (2014). "Alveolar macrophages: plasticity in a tissue-specific context." *Nat Rev Immunol* 14(2): 81-93.

Janson, C. (2004). "The effect of passive smoking on respiratory health in children and adults." *Int J Tuberc Lung Dis* 8(5): 510-516.

Klepeis, N. E., W. C. Nelson, W. R. Ott, J. P. Robinson, A. M. Tsang, P. Switzer, J. V. Behar, S. C. Hern and W. H. Engelmann (2001). "The National Human Activity Pattern Survey (NHAPS): a resource for assessing exposure to environmental pollutants." *J Expo Anal Environ Epidemiol* 11(3): 231-252.

Knorr, B., L. M. Franchi, H. Bisgaard, J. H. Vermeulen, P. LeSouef, N. Santanello, T. M. Michele, T. F. Reiss, H. H. Nguyen and D. L. Bratton (2001). "Montelukast, a leukotriene receptor antagonist, for the treatment of persistent asthma in children aged 2 to 5 years." *Pediatrics* 108(3): E48.

Kohler, E., V. Sollich, R. Schuster and W. Thal (1999). "Passive smoke exposure in infants and children with respiratory tract diseases." *Hum Exp Toxicol* 18(4): 212-217.

Kopp, B. T., L. Sarzynski, S. Khalfoun, D. Hayes, Jr., R. Thompson, L. Nicholson, F. Long, R. Castile and J. Groner (2015). "Detrimental effects of secondhand smoke exposure on infants with cystic fibrosis." *Pediatr Pulmonol* 50(1): 25-34.

Kostikas, K., M. Minas, E. Nikolaou, A. I. Papaioannou, P. Liakos, S. Gougoura, K. I. Gourgoulianis, P. C. Dinas, G. S. Metsios, A. Z. Jamurtas, A. D. Flouris and Y. Koutedakis (2013). "Secondhand smoke exposure induces acutely airway acidification and oxidative stress." *Respir Med* 107(2): 172-179.

Kovesi, T., M. Corey and H. Levison (1993). "Passive smoking and lung function in cystic fibrosis." *The American review of respiratory disease* 148(5): 1266-1271.

Laskin, D. L., V. R. Sunil, C. R. Gardner and J. D. Laskin (2011). "Macrophages and tissue injury: agents of defense or destruction?" *Annu Rev Pharmacol Toxicol* 51: 267-288.

Lazarus, S. C., V. M. Chinchilli, N. J. Rollings, H. A. Boushey, R. Cherniack, T. J. Craig, A. Deykin, E. DiMango, J. E. Fish, J. G. Ford, E. Israel, J. Kiley, M. Kraft, R. F. Lemanske, Jr., F. T. Leone, R. J. Martin, G. R. Pesola, S. P. Peters, C. A. Sorkness, S. J. Szefler, M. E. Wechsler, J. V. Fahy, L. National Heart and N. Blood Institute's Asthma Clinical Research (2007). "Smoking affects response to inhaled corticosteroids or leukotriene receptor antagonists in asthma." *Am J Respir Crit Care Med* 175(8): 783-790.

Leopold, P. L., M. J. O'Mahony, X. J. Lian, A. E. Tilley, B. G. Harvey and R. G. Crystal (2009). "Smoking is associated with shortened airway cilia." *PLoS One* 4(12): e8157.

Liu, Y. and Y. P. Di (2012). "Effects of second hand smoke on airway secretion and mucociliary clearance." *Front Physiol* 3: 342.

Livraghi, A. and S. H. Randell (2007). "Cystic fibrosis and other respiratory diseases of impaired mucus clearance." *Toxicol Pathol* 35(1): 116-129.

Martin, T. R. and C. W. Frevert (2005). "Innate immunity in the lungs." *Proc Am Thorac Soc* 2(5): 403-411.

Masoli, M., D. Fabian, S. Holt, R. Beasley and P. Global Initiative for Asthma (2004). "The global burden of asthma: executive summary of the GINA Dissemination Committee report." *Allergy* 59(5): 469-478.

Mills, C. D. (2012). "M1 and M2 Macrophages: Oracles of Health and Disease." *Crit Rev Immunol* 32(6): 463-488.

Moritsugu, K. P. (2007). "The 2006 Report of the Surgeon General: the health consequences of involuntary exposure to tobacco smoke." *American journal of preventive medicine* 32(6): 542-543.

Munkholm, M. and J. Mortensen (2014). "Mucociliary clearance: pathophysiological aspects." *Clin Physiol Funct Imaging* 34(3): 171-177.

Navarrette, C. R., J. H. Sisson, E. Nance, D. Allen-Gipson, J. Hanes and T. A. Wyatt (2012). "Particulate matter in cigarette smoke increases ciliary axoneme beating through mechanical stimulation." *J Aerosol Med Pulm Drug Deliv* 25(3): 159-168.

Neuman, A., C. Hohmann, N. Orsini, G. Pershagen, E. Eller, H. F. Kjaer, U. Gehring, R. Granell, J. Henderson, J. Heinrich, S. Lau, M. Nieuwenhuijsen, J. Sunyer, C. Tischer, M. Torrent, U. Wahn, A. H. Wijga, M. Wickman, T. Keil, A. Bergstrom and E. Consortium (2012). "Maternal smoking in pregnancy and asthma in preschool children: a pooled analysis of eight birth cohorts." *Am J Respir Crit Care Med* 186(10): 1037-1043.

Olin, J. T. and M. E. Wechsler (2014). "Asthma: pathogenesis and novel drugs for treatment." *BMJ 3*49: g5517.

Ortega-Garcia, J. A., M. T. Lopez-Fernandez, R. Llano, M. D. Pastor-Vivero, P. Mondejar-Lopez, M. F. Sanchez-Sauco and M. Sanchez-Solis (2012). "Smoking prevention and cessation programme in cystic fibrosis: integrating an environmental health approach." *Journal of cystic fibrosis : official journal of the European Cystic Fibrosis Society* 11(1): 34-39.

Rabinovitch, N., N. Reisdorph, L. Silveira and E. W. Gelfand (2011). "Urinary leukotriene E(4) levels identify children with tobacco smoke exposure at risk for asthma exacerbation." *J Allergy Clin Immunol* 128(2): 323-327.

Rabinovitch, N., M. Strand, K. Stuhlman and E. W. Gelfand (2008). "Exposure to tobacco smoke increases leukotriene E4-related albuterol usage and response to montelukast." *J Allergy Clin Immunol* 121(6): 1365-1371.

Raju, S. V., P. L. Jackson, C. A. Courville, C. M. McNicholas, P. A. Sloane, G. Sabbatini, S. Tidwell, L. P. Tang, B. Liu, J. A. Fortenberry, C. W. Jones, J. A. Boydston, J. P. Clancy, L. E. Bowen, F. J. Accurso, J. E. Blalock, M. T. Dransfield and S. M. Rowe (2013). "Cigarette smoke induces systemic defects in cystic fibrosis transmembrane conductance regulator function." *Am J Respir Crit Care Med* 188(11): 1321-1330.

Rasmussen, J. E., J. T. Sheridan, W. Polk, C. M. Davies and R. Tarran (2014). "Cigarette smoke-induced Ca2+ release leads to cystic fibrosis transmembrane conductance regulator (CFTR) dysfunction." *J Biol Chem* 289(11): 7671-7681.

Ratjen, F. (2006). "Restoring airway surface liquid in cystic fibrosis." *N Engl J Med* 354(3): 291-293.

Rosenfeld, M., P. M. Farrell, M. Kloster, J. O. Swanson, T. Vu, L. Brumback, J. D. Acton, R. G. Castile, A. A. Colin, C. K. Conrad, M. A. Hart, G. S.

Kerby, P. W. Hiatt, P. J. Mogayzel, R. C. Johnson and S. D. Davis (2013). "Association of lung function, chest radiographs and clinical features in infants with cystic fibrosis." *The European respiratory journal* 42(6): 1545-1552.

Rubin, B. K. (1990). "Exposure of children with cystic fibrosis to environmental tobacco smoke." *N Engl J Med* 323(12): 782-788.

Schvartsman, C., S. C. Farhat, S. Schvartsman and P. H. Saldiva (2013). "Parental smoking patterns and their association with wheezing in children." *Clinics (Sao Paulo)* 68(7): 934-939.

Shaykhiev, R., F. Otaki, P. Bonsu, D. T. Dang, M. Teater, Y. Strulovici-Barel, J. Salit, B. G. Harvey and R. G. Crystal (2011). "Cigarette smoking reprograms apical junctional complex molecular architecture in the human airway epithelium in vivo." *Cell Mol Life Sci* 68(5): 877-892.

Smyth, A., U. O'Hea, G. Williams, R. Smyth and D. Heaf (1994). "Passive smoking and impaired lung function in cystic fibrosis." *Arch Dis Child* 71(4): 353-354.

Sopori, M. (2002). "Effects of cigarette smoke on the immune system." *Nat Rev Immunol* 2(5): 372-377.

Stampfli, M. R. and G. P. Anderson (2009). "How cigarette smoke skews immune responses to promote infection, lung disease and cancer." *Nat Rev Immunol* 9(5): 377-384.

Strachan, D. P. and D. G. Cook (1997). "Health effects of passive smoking. 1. Parental smoking and lower respiratory illness in infancy and early childhood." *Thorax* 52(10): 905-914.

Strausbaugh, S. D. and P. B. Davis (2007). "Cystic fibrosis: a review of epidemiology and pathobiology." *Clin Chest Med* 28(2): 279-288.

Sun, Y., S. Ito, N. Nishio, Y. Tanaka, N. Chen and K. Isobe (2014). "Acrolein induced both pulmonary inflammation and the death of lung epithelial cells." *Toxicol Lett* 229(2): 384-392.

Tai, A., H. Tran, M. Roberts, N. Clarke, A. M. Gibson, S. Vidmar, J. Wilson and C. F. Robertson (2014). "Outcomes of childhood asthma to the age of 50 years." *J Allergy Clin Immunol* 133(6): 1572-1578 e1573.

Tilley, A. E., M. S. Walters, R. Shaykhiev and R. G. Crystal (2015). "Cilia dysfunction in lung disease." *Annu Rev Physiol* 77: 379-406.

Valsamis, C., S. Krishnan and A. J. Dozor (2014). "The effects of low-level environmental tobacco smoke exposure on pulmonary function tests in preschool children with asthma." *J Asthma* 51(7): 685-690.

Verma, A., D. Clough, D. McKenna, M. Dodd and A. K. Webb (2001). "Smoking and cystic fibrosis." *J R Soc Med* 94 Suppl 40: 29-34.

Wall, M. (2000). "Pre- and postnatal smoking intervention in managed care settings." *Tob Control* 9 Suppl 1: I63.

Winickoff, J. P., A. B. Berkowitz, K. Brooks, S. E. Tanski, A. Geller, C. Thomson, H. A. Lando, S. Curry, M. Muramoto, A. V. Prokhorov, D. Best, M. Weitzman, L. Pbert and C. f. C. H. R. o. t. A. A. o. P. Tobacco Consortium (2005). "State-of-the-art interventions for office-based parental tobacco control." *Pediatrics* 115(3): 750-760.

Xiao, C., S. M. Puddicombe, S. Field, J. Haywood, V. Broughton-Head, I. Puxeddu, H. M. Haitchi, E. Vernon-Wilson, D. Sammut, N. Bedke, C. Cremin, J. Sones, R. Djukanovic, P. H. Howarth, J. E. Collins, S. T. Holgate, P. Monk and D. E. Davies (2011). "Defective epithelial barrier function in asthma." *J Allergy Clin Immunol* 128(3): 549-556 e541-512.

Zhou, H., X. Wang, L. Brighton, M. Hazucha, I. Jaspers and J. L. Carson (2009). "Increased nasal epithelial ciliary beat frequency associated with lifestyle tobacco smoke exposure." *Inhal Toxicol* 21(10): 875-881.

- 18 mg/ml: High volume smoker, more than a pack per day
- 24 mg/ml: Heavy smoker, this is typically the highest strength sold by any reputable retailer

Nicotine which is used is derived from tobacco plants or tobacco dust and it is contained in the liquid as natural or synthetic form. [19]

However nicotine inhalation can be affected by several factors. These are the quality and construction characteristics of each brand, concentration of nicotine in e-liquids and user's puffing behavior (length and frequency of puffs and depth of inhalation). This means that a faster and deeper puff increases nicotine delivery and fulfils the needs of the vaper. [20]

A study in the UK among five brands with high internet popularity shows that the amount of nicotine which is dissolved in the e-liquid was not related to the amount of nicotine in the steam. This demonstrates that the technical characteristics of each brand determinate the nicotine delivery. [12]

Another extensive report indicates that there are cases when the amount of nicotine contained in the liquid does not correspond to the value which is labeled and be highly fluctuated with some samples contain 100mg/ml. [15]

Microchip-Wick

Wick is a part of the cartomizer tank and its use is to direct the e-juice from the tank to the heating element of atomizer. [7]

Atomizer or Vaporization Chamber

The atomizer is the heating element of e-cig which vaporizes the e-liquid. It consists of two parts, a battery- powered heating unit which is activated from puffing and a reservoir which draws the liquid and drains it to the heating element. [18] The latter consists of a metal wire which is wound around the tank draining the liquid, creating a coil and is connected to the positive and negative terminals of the battery. The coil operates as a resistor-microchip which is heated quickly. Upon heating, the liquid evaporates and produces an aerosol that simulates the smoke of a real cigarette and then it is inhaled by the user. [21, 22, 23, 24] The resistance of the atomizer varies from 1.5 Ohm to 3 Ohm. The smaller the resistance, the higher the steam which is produced and the nicotine which is inhaled by the user. [25] Silica fibers are often used for the construction of the atomizer. [26] In rebuildable or MODs e-cig the vapers themselves can construct this part of e-cig and select any desired resistance. [8, 11] Usually the rebuildable e-cigs have lower resistances. [18] In these the atomizer can be constructed also by other materials such as oxidized stainless steel, porous ceramic, and cotton or bamboo yarn. [25]

Battery

The battery is the electrical source that powers the e-cig to generate the atomizer and vaporize the e-liquid. [18] Battery power depends on two different measures, voltage (v) and mill ampere (mAh). Most electronic cigarettes use 3.7 volt batteries but operate with different capacities (mAh). Although all 3.7v batteries have the same current (v), they all have different capacities (mAh). The higher the mAh number the longer they will last and be able to generate vapor. [25]

Red LED

The red LED is the indicator light in the end of the e-cig and simulates a burning cigarette tip [17]. In some models this LED is powered from the activation of the heating element while in manually operated models is powered by pressing a button. [5]

Generations of Electronic Cigarette

As of 2014 there have been three prevalent types of devices available.

First generation e-cigs resemble a conventional tobacco cigarette in the look and the feel, [9] thus called "cigalikes" [8, 14] and consist of a small low power lithium battery and a prefilled cartomizer-tank with liquid which is used from the atomizer. When the battery dies or the liquid is used up the device is disposed. The battery could be rechargeable or used only once. When charging the battery, depending on the model, the vaper may use the device or not. In some models the electronic sensor is activated by drawing breath through the device or in others by pressing a button. The LED in the button or the end glows when the device is in operation. [27]

Second generation e-cigs consist a higher-capacity and power lithium battery and a cartomizer-tank which can be refilled with e-liquid (e-juice). This liquid blended with nicotine and flavors is sold in separate bottles. Similarly to the first generation there is a button which activates the battery. Battery can be charged also with USB charger. [5]

Third-generation devices consists of very large capacity and rechargeable batteries and rebuildable atomizers with replacement heating coils [14] that allow vapers to change the voltage and wattage. The resistance may be single horizontal or double vertical, fluctuating the amount of smoke produced and thus of nicotine derived. Third generation cigarettes give the ability the users to prepare and regulate the amount of liquid which they want to bath the atomizer and be vaporized. These devices are also called 'Mods' from modifications, PVs (personal vaporizer) or APVs (advanced personal vaporizer). The size of the battery allows the use of larger tanks which can contain a larger quantity e - liquid. [5, 28]

The main part of e-cig is usually made of aluminum and the electrical parts are manufactured of stainless steel or brass. The materials aim to use the full power of battery improving the quantity and quality of the mist which is produced. [5, 8, 28]

Regarding the third generation of-cigs, the user can modify and effectively assemble the device himself. [28] There are cases when the user modifies the e-liquid and instead of nicotine he/she adds marijuana. Also, experts even say that marijuana does not emit the apocalyptic smell when smoked in an e-cig. The possibility of modifying the device can lead to vape uncontrolled nicotine levels or other substances and generate suspicious steam. [29]

Recently, some e-cigs companies hale the fourth generation e-cig which contains computer chips to control steam temperature, inhalation volume control, lights when another steamer is near and mimicking the form of everyday items such as pen and USB memory sticks. [30, 31]

Nowadays, a new model of e-cig is advertised; e-cig with Bluetooth phone function that allow the vaper to make and receive calls via Bluetooth and play music via the built-in microphone. [31]

However nicotine inhalation can be affected by several factors. These are the quality and construction characteristics of each brand, concentration of nicotine in e-liquids and user's puffing behavior (length and frequency of puffs and depth of inhalation). This means that a faster and deeper puff increases nicotine delivery and fulfils the needs of the vaper.

LEGISLATION FOR ELECTRONIC CIGARETTE

The promotion of electronic nicotine delivery systems as substitutes for smoking conventional tobacco products has rapidly increased the sales of electronic cigarettes worldwide. One only Google search for e-cig has increased 5000%over in 2010-2011. [32] Alongside, the intense research into the effects of-cigs has brought to light a growing body of evidence that confirms the negative effects of electronic cigarette smoking both on smokers (first hand smoking) and the general adult and pediatric population (second and third hand smoke). [6, 33, 34] This reality raises concerns about the health of the global community and pushes the state authorities to adopt regulatory measures for vaping. [34]

According to the WHO, before governments design any regulatory strategy for ENDs, they should take into consideration some general instructions. These are that ENDs should not be promoted to non-smokers, pregnant women and youth, governments should diminish any health risks for users and public, prohibit unsubstantiated health claims about the use of e-cigs and protect the health of citizens from the commercial and other vested interests of the tobacco industry. [35]

Ideally the competent authorities, before any regulations about the prohibition or not of e-cigs be enacted, should take into account all the scientific results of the researches and studies on e-cigs, examine all possible solutions and choose the best for the health and safety of the global community. [34] The lack of data regarding the safety of e-cigs demonstrates how important and necessary research in this field is. [7, 34]

In international law there is a principle called precautionary (PP) which claims that prevention is better than cure. The precautionary principle applies when a process, phenomenon or product displays potentially dangerous effects without the scientific and objective consent. Based on this principle many anti-tobacco organizations have suggested restrictive regulations. [36]

The first report by the World Health Organization (WHO) Study Group on Tobacco Product Regulation suggests a precautionary approach, for the most part, because the evidence about the safety of e-cigs and their contribution on smoking cessation was virtually nonexistent at that time. [37]

This report also stated that is more research on benefits and potential harms of e-cigarettes must be conducted. [37] Today, several scientific opinions diverge with some supporting the safe use of e-cig and its liquid as a smoking cessation method and others raising serious questions on it. [38]

The relative easy access to youth and non-smokers in electronic nicotine delivery systems [44] and the adverse effects reports to the general population and pediatric population concern Food and Drug Administration (FDA) and Family Smoking Prevention and Tobacco Control Act (TCA). [6, 33, 39]

The Center for Tobacco Products (CTP) applying the Tobacco Control Act (TCA) comes into contact with consumers, health professionals and all members of the public interested in e-cigs. This center considers the side effects of e-cig, checks the safety of products and takes appropriate measures for user education. [38, 40]

United States

In 2009 the US FDA undertook the regulation of tobacco products and the power to set conditions of production, marketing and distribution of electronic cigarettes. [41]

The FDA classifies them in the category of pharmaceutical products as drug delivery devices. Therefore, they are regulated by the Food, Drug and Cosmetic Act (FDCA). This categorization changed by court decision in January 2010 when the Federal District Court Judge Richard J. Leon decided ENDs join the same rules relating to tobacco products unless placed on the market for proven therapeutic reasons and so regulated as drugs or devices. [42, 43]

Currently, e-cigarettes that are marketed for therapeutic purposes are regulated by the FDA Center for Drug Evaluation and Research (FDA. (2012). [44]

In April 2014 the FDA released a draft of the deeming rule. According to this the electronic cigarettes' manufacturers should:

1) Register in FDA and report components
2) Require childproof caps on ENDS liquids
3) include warning labels on products and their advertisements
4) Promote to the market only products that have been reviewed by the FDA and prohibit youth-oriented marketing and sales
5) Do not distribute free samples
6) Limit the sale of e-cigs strictly to adults (18 and older)
7) Prohibit the use of e-cigs in places where cigarette smoke is prohibited [39, 45]

This plan does not limit the range of flavors in e-liquid nor online sales. The period of this project ended in August 2014 and the final action is expected in June 2015. [41]

The absence of FDA tobacco regulatory authority instructions of FDA creates a lack of compulsory standards for the construction of e-cigs and few quality controls. [46]

The lack of regulatory measures on flavorings and advertising of e-cigs concerns global community since at the time of publication, there are over 470 brands and 7700 flavors [47] and youth exposures to television e - cigarettes advertisements increased 256 % over 2011-2013 while young adults exposure increased 321 % over the same period. [48]

Moreover, there are hundreds of brands of e-cigs on the market and the results of quality test vary not only between brands but also among models of the same manufacturer. A research has found misleading information about the components of the e-cig in many cartomizer. E-cigs labeled as nicotine free, in fact contains nicotine and vice versa. Also there are complaints of leaky cartomizers and clearomizers and other defective parts.

The legal vacuum that exists on the issue of e-cigs has prompted many states to adopt their own regulations. [49]

Most states prohibit the sale of e-cigs to any person under the age of 18 and the use of them indoors to some degree. [49]

The FDA does not have the authority to tax ENDS or restrict their use in public spaces, though some states and local jurisdictions may do so.

Europe

Due to the rapid augmentation of the use of electronic nicotine delivery systems the European Commission on 19 December 2012 decided to adopt the proposal for the revision of the Directive of the European Union on tobacco products (2001/37 / EC), [50] which included proposals to introduce restrictions on the use and sales of electronic cigarettes. [51, 52]

On October 8, 2013, the European Parliament in Strasbourg voted against the Commission's proposal to introduce e-cigs as medical pharmaceutical preparations, but suggested that cross-border marketing of electronic cigarettes be regulated similarly to tobacco products, which means that in the European Union the sale of them to people under 18 is prohibited. Warning labels will also be required.

Five months later (February 2014), the European Parliament approved new regulations for tobacco products which include-cigs.

The new regulations prohibit the advertising of e-cigs, set limits on maximum concentrations of nicotine liquid, limiting the maximum amount of liquid that can be sold, require safety caps for children and jelly packaging of liquid. These regulations place demands on the purity of the ingredients, require that electronic cigarettes devices to produce stable vapor doses and require the brands to indicate on the packaging the ingredients of e-liquids and the amount of soluted nicotine. The manufacturers are obliged to place health warnings about the addiction and the toxicity of nicotine and instructions for the use of e-cigs. Any elements or characteristics which promote ENDs and vaping are prohibited.

This directive increases the power of regulator authorities to act when the above regulations are violated.

The new Directive came into force in May 2014. The European Parliament has left an adjustment period of two years to all Member States to adapt their national legislation in accordance with the revised Directive and further to allow manufacturers to market their existing stock, if this is compatible with the old Directive or other relevant legislation. The new rules in respect of any state will apply in the first half of 2016 and benefit both consumers and e-cig manufacturers. [53]

Until today, in Europe, some countries consider the nicotine containing cartomizer as medical products and e-cigs as medical devices (Austria, Denmark), some allow the sale and use of e-cigs devices and nicotine cartomizers (Germany, Ireland), others allow their use but not the sale of nicotine chambers -so the users should purchase them from other EEA member states (Norway, Hungary, Portugal) while the Netherlands bans completely the sale and advertising of e-cigs. [22]

Other Countries

In Australia, the authorities federally prohibit the sale of e-cigs with nicotine but people can legally source them from other countries. It is necessary the devices' manufacturers do not claim that e-cigs act therapeutically. However, some regions such as WA, Qld and SA, the sale of free-nicotine also appears to be prohibited. [54, 55]

In Canada, Vancouver city included e-cigs in tobacco products and therefore its use is prohibited in all public places. The same regulation is adopted by the municipality of York in Ontario and the city of Red Deer in Alberta. [56]

Brazil, Panama, Singapore and Lebanon banned the distribution and the importation of e-cig due to the lack of sufficient evidence about the safety of these products. [57]

POSITION OF WORLD HEALTH ORGANIZATION (WHO) ON THE ISSUE OF ELECTRONIC CIGARETTES

In July 2014, the World Health Organization (WHO) prepared a report for the Conference of the Parties to the WHO Framework Convention on Tobacco Control. In this report, WHO mentioned that more research and evidence are needed in order to decide if e-cigs help people quit smoking. It suggested the countries to adopt regulations about the use, the sale and the advertising of ENDs, users to be careful about the brands and the origin of the devices and finally report the health risks of e-cigs to the vapers and the non-users. [35]

Health

In the general population there is a perception that Electronic Nicotine Delivery Systems are healthy substitutes of tobacco cigarettes. This has occurred due to the aggressive marketing, the increased strict regulations of conventional cigarettes and the ignorance of the possible damages of e-cigs. [47]

An internet survey including 81 participants showed the two mains reason for starting using e-cig is to quit smoking or try a safer alternative way of smoking. The majority of participants (85%) believe that e-cig is not absolutely healthy but it is less dangerous than conventional cigarette. The 80% of them stopped smoking when they started vaping. [47]

The first clinical study in the world was conducted in 2012 by a Greek team headed by Dr. Konstantinos Farsalinos from Onassis Cardiac Surgery Center in Athens. This team examined the heart and lung function of twenty young smokers before and after smoking conventional tobacco cigarette against that of twenty-two e-cig users before and after using the device for seven minutes. The results showed that vaping did not cause adverse cardiac effects but it caused little irritation of lungs and decrease of pulmonary resistances. The conclusions of this study are that ENDs are far less better than conventional cigarettes because they do not produce dangerous chemicals of the tobacco's burn (tar, carbon monoxide and etc) and it is necessary for bigger and long-term studies to be conducted to examine the effects of e-cigs. [48]

From 2012 to today the studies about health effects on humans are few and limited and in vivo animal models have not been generated. [47]

Toxic Substances in the Smoke of E-Cig

The conventional cigarettes smoke contains over 4000 chemical compounds which cause damages and are produced by the tobacco combustion at temperatures 500-800 degrees. E-cig is not based on burning but on liquid evaporation at temperatures below 200 degrees. This implies that the vapor does not contain tar, carbon monoxide, hydrocyanic acid, arsenic, lead and mercury which are carcinogenic.

The principle of produced vapor from Propylene Glycol is based on the one that has been used by mist generators. Studies that conducted in order to examine if PG produced harms Broadway's actors indicated that the excessive and long exposure irritates the respiratory system. [49] The study of University of British Columbia in Canada found that PG aerosol is responsible for eye and respiratory irritation and impairs the lung function. [50] Nevertheless the FDA categorized PG as safe to produce vapor. [6] Glycerol is the second substance which is used in e-liquid. It is widely used in foods and chemicals and it is considered nontoxic. The high temperature in the nicotine chamber may cause it to produce the toxic acrolein. [5] The flavors which contain the e-cig have been used for many years. However, there is no information about the short and long term impacts on humans from their inhalation. [6] The conventional cigarettes produce two classes of radicals during the process of combustion. The gas phase radicals have a short self life and so their effects on health have not been determined yet. Instead, TPM have long life and ability to settle in tissues with which they came into contact. [51]

A recent survey found that the vapour of e-cig contains a remarkable number of free radicals (7x1011 free radicals per puff) that could be potentially toxic to cells. [47]

Research conducted in October 2014 studied whether the vapor of e-cigs contains toxic and carcinogenic carbonyl compounds. Under normal operating conditions (battery voltage 3, 2-4, 8) quantities of formaldehyde and acetaldehyde were found in 8 of the 13 samples, in lower levels of course from the tobacco smoke. Remarkably, the same study showed that the increase of the battery voltage from 3.2 to 4.8 means 200-fold higher levels of formaldehyde acetaldehyde and acetone in the vapor of e-cigs. Indeed levels of formaldehyde vapor in the high-voltage batteries approached those of conventional cigarettes. This research concluded that the vapor of e-cigs contains toxic and carcinogenic carbonyl compounds and the high-voltage devices may expose the vapers to dangerous levels of them. [52]

A recent survey which was published in New England Journal of Medicine reported that the atomizer of e-cigs with high voltage battery (5Volt) produced a new kind formaldehyde which had not been identified before. This substance is extremely toxic and makes the risk of cancer five to fifteen times higher than that of conventional cigarettes. [53]

A concern which recently has been raised is that cartomizers may be filled with liquid in which the users have soluted other drugs such as marihuana. [22]

Moreover, in many cases e-liquids have been implicated for increasing the number of children and adults poisoned. [27]

Researchers at the University of California claimed that the distribution and use of e-cigs should be banned because the toxicity levels of the chemical propylene glycol are still unknown and there is a hazard the users substitute an addiction for another, just as harmful too. [54]

Impacts on the Human Organization

Aggravating circumstances that e-cig weakens the immune system by reducing phagocytosis by alveolar macrophages and increases significantly the oxidative stress and more reduces the body's defenses to bacterial and viral agents were recorded after exposure of mice to e-cig vapor containing nicotine comparable to e -cig human users for 2 weeks. [55]

It is known that regular tobacco smoke has serious consequences on cells that line the airway predisposing them to become lung cancer. The Professor of Medicine in Boston University School of Medicine Director and his team have similar concerns about e-cigs after what they have seen so far is when they take those cells, put them in a dish and expose them to e-cig smoke; some negative changes are induced in the genes. Genes that are changed with regular tobacco smoke are changed in a similar matter with e-cig. [56]

An additional concern is the exposure to excessive and uncontrolled amounts of nicotine. The advents reports in United States and United Kingdom

[14] Davis B, Dang M, Kim J, Talbot P.Nicotine concentrations in electronic cigarette refill and do-it-yourself fluids. *Nicotine Tob Res.* 2015 Feb;17(2):134-41.

[15] Williams M, Villarreal A, Bozhilov K, Lin S, Talbot P. Metal and silicate particles including nanoparticles are present in electronic cigarette cartomizer fluid and aerosol. *PLoS One.* 2013;8(3):e57987.

[16] Bertholon JF, Becquemin MH, Annesi-Maesano I, Dautzenberg B. Electronic cigarettes: a short review. *Respiration.* 2013;86(5):433-8.

[17] McQueen A, Tower S, Sumner W. Interviews with "vapers": implications for future research with electronic cigarettes. *Nicotine Tob Res.* 2011 Sep;13(9):860-7.

[18] Werley MS, McDonald P, Lilly P, Kirkpatrick D, Wallery J, Byron P et al. Non-clinical safety and pharmacokinetic evaluations of propylene glycol aerosol in Sprague-Dawley rats and Beagle dogs. *Toxicology.* 2011 Sep 5;287(1-3):76-90.

[19] Chang H. Research gaps related to the environmental impacts of electronic cigarettes. *Tob Control.* 2014 May;23 Suppl 2:ii54-8.

[20] WHO Framework Convention on Tobacco Control. Electronic nicotine delivery systems. *Proceedings of the 6th session Conference of the Parties to the WHO Framework Convention on Tobacco Control;* 2014 Oct13-18; Moscow, Russian.

[21] Ngo N. As e-cigarette stores pop up in Twin Cities, so do the questions. Pioneer Press [Internet] 2013 Sept 28. [cited 2013 Sept 28]; Available from: http://www.twincities.com/health/ci_24192074/e-cigarette-stores-pop-up-twin-cities-so.

[22] Givens A, Cheng PS. I-Team: E-Cigarettes, Used to Smoke Marijuana, Spark New Concerns. NBC 4 New York . [Internet] 2013 Oct 10. [cited 2013 Oct 10]; Available from: http://www.nbcnewyork.com/news/local /ECigarettes-Drugs-Marijuana-Vapor-Pens-Smoking-I-Team-227269001.html.

[23] LeClair D. Smokio Smart e-cigarette allows users to monitor their habit via smartphone app. GizMag [Internet] 2013 Dec 19. [cited 2013 Dec 19]; Available from: http://www.gizmag.com/smokio-smart-e-cigarettes /30193/.

[24] Yamin CK, Bitton A, Bates DW. E-cigarettes: a rapidly growing Internet phenomenon. *Ann Intern Med.* 2010 Nov 2;153(9):607-9.

[25] Mc Robbie H. Electronic cigarettes. England: National Centre for Smoking Cessation and Training; 2014.

[26] Grothaus M. Trading addictions: the inside story of the e-cig modding scene. Engadget. [Internet] 2014 Oct 01. [cited 2014 Oct 01]; Available from: http://www.engadget.com/2014/10/01/inside-story-e-cig-modding-uk/.

[27] Durmowicz EL. The impact of electronic cigarettes on the paediatric population. *Tob Control.* 2014 May;23 Suppl 2:ii41-6.

[28] Saitta D, Ferro GA, Polosa R. Achieving appropriate regulations for electronic cigarettes. *Ther Adv Chronic Dis.* 2014 Mar;5(2):50-61.

[29] Grandjean P. Implications of the precautionary principle for primary prevention and research. *Annu Rev Public Health*. 2004;25:199-223.

[30] WHO Study Group on Tobacco Product Regulation. Report on the scientific basis of tobacco product regulation: third report of a WHO Study Group. World Health Organization technical report series. 2009(955):1-41, back cover.

[31] Chen IL. FDA summary of adverse events on electronic cigarettes. *Nicotine Tob Res.* 2013 Feb;15(2):615-6.

[32] Brandon TH, Goniewicz ML, Hanna NH, Hatsukami DK, Herbst RS, Hobin JA, et al. Electronic nicotine delivery systems: a policy statement from the american association for cancer research and the american society of clinical oncology. *Clin Cancer Res.* 2015 Feb 1;21(3):514-25.

[33] Riker CA, Lee K, Darville A, Hahn EJ. E-cigarettes: promise or peril? *Nurs Clin North Am.* 2012 Mar;47(1):159-71.

[34] King BA, Alam S, Promoff G, Arrazola R, Dube SR. Awareness and ever-use of electronic cigarettes among U.S. adults, 2010-2011. *Nicotine Tob Res.* 2013 Sep;15(9):1623-7.

[35] Wilson D. Judge Orders F.D.A. to Stop Blocking Imports of E-Cigarettes From China. The New York Times. [Internet] 2010 Jan 14. [cited 2010 Jan 14]; Available from: http://www.nytimes.com/2010/01/15/business/15smoke.html?_r=0.

[36] Sottera, Inc. v. Food & Drug Admin., 627 F.3d 891, 393 U.S.App.D.C. 257 (D.C. Cir., 2011). Available from: http://www.tobaccocontrollaws.org/files/live/litigation/1032/US_Sottera%20v.%20Food%20%26%20Drug%20Adminis.pdf.

[37] American Non-Smokers' Rights Foundation. U.S. and Local Laws Regulating Use of E-Cigarettes. Available from: http://www.no-smoke.org/pdf/ecigslaws.pdf.

[38] Benowitz NL, Goniewicz ML. The regulatory challenge of electronic cigarettes. *JAMA.* 2013 Aug 21;310(7):685-6.

[39] Zhu SH, Sun JY, Bonnevie E, Cummins SE, Gamst A, Yin L, et al. Four hundred and sixty brands of e-cigarettes and counting: implications for product regulation. *Tob Control.* 2014 Jul;23 Suppl 3:iii3-9.

[40] Duke JC, Lee YO, Kim AE, Watson KA, Arnold KY, Nonnemaker JM, et al. Exposure to electronic cigarette television advertisements among youth and young adults. *Pediatrics.* 2014 Jul;134(1):e29-36.

[41] University of Maryland Law School. Electronic Cigarette Legislation Prohibiting Sale to Minors in Other States [Internet} 2013 Aug 30. Available from: http://www.law.umaryland.edu/programs/tobacco /documents/E-Cig_Legislation.pdf.

[42] European Parliament and the Council of European Union. Tobacco Products Directive (2014/40/EU). 2014 Apr 3; Available from: http://ec.europa.eu/health/tobacco/docs/dir_201440_en.pdf.

[43] Fraser C. EU plans tougher tobacco restrictions on e-cigarettes. BBC News Online. [Internet] 2013 Oct 8. [cited 2013 Oct 8]; Available from: http://www.bbc.com/news/world-europe-24440091.

[44] European Commission. Questions & Answers: New rules for tobacco products. Brussels;2014. Available from: http://europa.eu/rapid/press-release_MEMO-14-134_en.htm.

[45] Carrick D. Science and the law hazy on e-cigarettes. Abc net. [Internet]. 2014 Jun 12. [cited 2014 Jun 12]; Available from: Morton B. Vancouver to treat e-cigarettes like tobacco. The Vancouver Sun. [Internet] 2014 Oct 2.[cited 2014 Oct 2]; Available from: http://www.vancouversun.com /news/Vancouver+treat+cigarettes+like+tobacco/10257540/story.html.

[46] Goniewicz ML, Lingas EO, Hajek P. Patterns of electronic cigarette use and user beliefs about their safety and benefits: an internet survey. *Drug Alcohol Rev.* 2013 Mar;32(2):133-40.

[47] Hirschler B. Greek study finds e-cigarettes no threat to heart. Reuters. [Internet] 2012 Aug 25. [cited 2012 Aug 25]; Available from: http://www.reuters.com/article/2012/08/25/us-heart-ecigarettes-idUSBR E87O05Y20120825.

[48] Burr GA, Van Gilder TJ, Trout DB, WilcoxTG, Driscoll R: NIOSH Health Hazard Evaluation HETA 90–0355–2499. Actor's Equity Association, 1994.

[49] Varughese S, Teschke K, Brauer M, Chow Y,van Netten C, Kennedy SM. Effects of theatrical smokes and fogs on respiratory health in the entertainment industry. *Am J Ind Med* 2005; 47: 411–418.

[50] Halliwell HB, Poulsen HE,editors. Cigarette Smoke and Oxidative Stress.New York: *Sringer;* 2006.

[51] Kosmider L, Sobczak A, Fik M, Knysak J, Zaciera M, Kurek J, et al. Carbonyl compounds in electronic cigarette vapors: effects of nicotine solvent and battery output voltage. *Nicotine Tob Res.* 2014 Oct;16(10):1319-26.

[52] Jensen RP1, Luo W, Pankow JF, Strongin RM, Peyton DH. Hidden formaldehyde in e-cigarette aerosols. *N Engl J Med.* 2015 Jan 22;372(4):392-4.

[53] Grana RA, Ling PM, Benowitz N, Glantz S. Electronic cigarettes. Cardiology patient page. *Circulation.* 2014 May 13;129(19):e490-2.

[54] Rimer S. Behind the vapor.Chasing the facts about e-cigarette health risks. Boston University Research. [Internet].2014 Oct 20. [cited 2014 Oct 20]; Available from: http://www.bu.edu/research/articles/behind-the-vapor/.

[55] Schripp T, Markewitz D, Uhde E, Salthammer T. Does e-cigarette consumption cause passive vaping? *Indoor Air.* 2013;23:25–31.

[56] Armitage AK, Dollery CT, George CF, Houseman TH, Lewis PJ, Turner DM. Absorption and metabolism of nicotine from cigarettes. *Br Med J.* 1975 Nov 8;4(5992):313-6.

[57] Czogala J, Goniewicz ML, Fidelus B, Zielinska-Danch W, Travers MJ, Sobczak A. Secondhand exposure to vapors from electronic cigarettes. *Nicotine Tob Res.* 2014 Jun;16(6):655-62.

[58] Schober W, Szendrei K, Matzen W, Osiander-Fuchs H, Heitmann D, Schettgen T et al. Use of electronic cigarettes (e-cigarettes) impairs indoor air quality and increases FeNO levels of e-cigarette consumers Int *J Hyg Environ Health.* 2014 Jul;217(6):628-37.

[59] Ellis M. Increased levels of toxic metals in e-cigarette secondhand smoke. Medical News Today.[Internet]. 2014 Sept 01. [cited2014 Sept 01]; Available from: http://www.medicalnewstoday.com/articles /281762.php.

[60] Ordonez JE, Kleinschmidt KC, Forrester MB. Electronic cigarette exposures reported to Texas poison centers. *Nicotine Tob Res.* 2015 Feb;17(2):209-11.

[61] Centers for Disease Control and Prevention. More than a quarter-million youth who had never smoked a cigarette used e-cigarettes in 2013. [Internet]. 2014 Aug 25; Available from: http://www.cdc.gov/media /releases/2014/p0825-e-cigarettes.html.

[62] Koch W. E-cigarettes: No smoke, but fiery debate over safety. USA today. [Internet]. 2012 Sept 17. [cited 2012 Sept 17]; Available from:

http://usatoday30.usatoday.com/news/health/story/2012-08-18/electronic-cigarettes-smokeless-vaping-risks/57121894/1.

[63] Cahn Z, Siegel M. Electronic cigarettes as a harm reduction strategy for tobacco control: a step forward or a repeat of past mistakes? *J Public Health Policy.* 2011 Feb;32(1):16-31.

[64] Farsalinos KE, Tsiapras D, Kyrzopoulos S, Savvopoulou M, Voudris V. Acute effects of using an electronic nicotine-delivery device (electronic cigarette) on myocardial function: comparison with the effects of regular cigarettes. *BMC Cardiovasc Disord.* 2014 Jun 23;14:78.

[65] Farsalinos KE, Romagna G. Chronic idiopathic neutrophilia in a smoker, relieved after smoking cessation with the use of electronic cigarette: a case report. *Clin Med Insights Case Rep.* 2013;6:15-21.

[66] McQueen A, Tower S, Sumner W. Interviews with "vapers": implications for future research with electronic cigarettes. Nicotine Tob Res. 2011 Sep;13(9):860-7.

[67] Polosa R, Caponnetto P, Morjaria JB, Papale G, Campagna D, Russo C. Effect of an electronic nicotine delivery device (e-cigarette) on smoking reduction and cessation: a prospective 6-month pilot study. *BMC Public Health.* 2011 Oct 11;11:786.

[68] Caponnetto P, Campagna D, Cibella F, Morjaria JB, Caruso M, Russo C et al. The efficacy and safety of an electronic cigarette (ECLAT) study: a prospective 12-month randomized control design study. *PLoS One.* 2013 Jun 24;8(6):e66317.

[69] Robinson D, Shubber K. European Commission puts money before health with e-cigarette tax proposals. *Financial Times.* 2015 Feb 18.

[70] Etter JF, Bullen C. A longitudinal study of electronic cigarette users. *Addict Behav.* 2014 Feb;39(2):491-4.

[71] Nicotine Science & Policy. E-cigarette taxation threat in Italy – expected to escalate cost of e-cigarettes [Internet]. 2014 Aug 18; Available from: http://nicotinepolicy.net/en/n-s-p/2354-e-cigarette-taxation-threat-in-italy-expected-to-escalate-cost-of-e-cigarettes.

[72] de Andrade M, Hastings G, Angus K.Promotion of electronic cigarettes: tobacco marketing reinvented? *BMJ.* 2013 Dec 21; 347():f7473.

[73] Grana RA, Glantz SA, Ling PM.Electronic nicotine delivery systems in the hands of Hollywood. *Tob Control.* 2011 Nov;20(6):425-6.

[74] Grana RA, Ling PM. "Smoking revolution": a content analysis of electronic cigarette retail websites. *Am J Prev Med.* 2014 Apr; 46(4):395-403.

[75] Kim AE, Lee YO, Shafer P, Nonnemaker J, Makarenko O. Adult smokers' receptivity to a television advert for electronic nicotine delivery systems. *Tob Control.* 2015 Mar;24(2):132-5.

In: Nicotine Dependence
Editor: Carlton L. Owens
ISBN: 978-1-63482-642-6

Chapter 7

SECONDHAND SMOKE: CHEMICAL COMPOSITION, EXPOSURE, HEALTH CONSEQUENCES AND PUBLIC POLICIES

Marcia Erazo[1*], Maria Paz Bertoglia[1], Juan Guillermo Gormaz[2], Carla Bertossi[1] and Matías Libuy[1]

[1]School of Public Health, Faculty of Medicine, University of Chile, Chile
[2]Molecular and Clinical Pharmacology Program, Institute of Biomedical Sciences, University of Chile, Chile

ABSTRACT

Secondhand smoke is the result of the stream produced by the burning cone in every puff, the exhaled smoke and the air that is in the room. The study of chemical composition of secondhand smoke has shown that is a complex mixture of constituents and the concentration of every compound depend on time and environmental conditions. There are different respirable suspended particles in secondhand smoke like

* Corresponding author: Marcia Erazo, Dr. Ph, School of Public Health, Faculty of Medicine, University of Chile, 1027 Independencia Avenue, P.O Box 70058, Santiago 7, Chile. Telephone: 56-2-29786813, E-mail:merazo@med.uchile.cl.

nicotine, carbon monoxide, benzene and formaldehyde, among others. Nicotine in hair is the most used biomarker in population-based studies.

Exposure to secondhand smoke takes place in indoor environments. Several studies in bar and restaurant's personnel have shown that there is a dose-response between indoor vapor-phase nicotine in bars or restaurants and the employee's hair nicotine concentration. This exposure has deleterious effects on the worker's health. Respiratory symptoms like wheeze, phlegm and throat pain have been described. Psychological effects are observed as well, including depressed mood, insomnia and anxiety. Furthermore, general population's health is also affected by secondhand smoke. In countries where a complete legislation ban to smoke in all indoor public places was implemented, a reduction in the prevalence rates of myocardial infarction after the implementation was reported.

According to the World Health Organization Framework Convention on Tobacco Control, countries that signed must implement comprehensive smoke-free legislation that prohibit tobacco smoking in all indoor public places, nevertheless, there are countries that still do not legislate, allowing to smoke in all areas, or have partial ban legislations, consenting to have mixed-areas inside a venue.

Keywords: Secondhand smoke, tobacco, effects

ABBREVIATIONS

SHS	secondhand smoke
ETS	environmental tobacco smoke
CO	carbon monoxide
WHO	World Health Organization
FCTC	Framework Convention on Tobacco Control
IARC	International Agency for Research in Cancer
HRQOL	Health related quality of life
DALYs	disability-adjusted life-years

1. INTRODUCTION

Secondhand smoke is a mixture of gases and fine particles that includes smoke from a burning tobacco product and smoke that has been exhaled or breathed out by the person or people smoking. This smoke contains almost

4,000 chemicals, including hundreds with known carcinogenic and noncarcinogenic health effects. The effects of these compounds are mediated through different mechanisms including direct irritant effects, immunological mechanisms, and mutagenesis [1-4] Consequently, there is a confirmed link between secondhand smoke exposure and disease, disability and death. Evidence about the links between second-hand smoke and specific diseases has been demonstrated in various types of epidemiological studies and it has been recognized as a risk factor for chronic respiratory symptoms and diseases, immunological pathology, cardiovascular disease, mental illness and different types of cancer.

Most exposure to secondhand smoke occurs in homes and workplaces. Secondhand smoke exposure also continues to occur in public places such as restaurants, bars, and casinos, as well as multiunit housing and vehicles. Comprehensive legislation to protect non-smokers from exposure to second-hand smoke in all indoor workplaces and public places has been implemented in some countries and subnational jurisdictions, but a major part of the world's population is still living without being protected by fully smoke-free public health regulations.

This chapter examines the chemical composition of secondhand smoke and it relation to harmful health effects, describes the conditions in which exposure to secondhand smoke occurs and finally, presents the current situation of smoke-free legislation in the world.

2. Chemical Composition of Secondhand Smoke

The secondhand smoke is the result of the stream produced by the burning cone in every puff, exhaled smoke and the air that is in the room. The study of chemical composition of secondhand smoke has shown that is a complex mixture of constituents and the concentration of every compound depend on time of exposure and environmental conditions. Almost 4000 constituents have been recognized in mainstream smoke, nevertheless, the qualitative composition of the smoke, vary.

There are different respirable suspended particles in secondhand smoke like nicotine, carbon monoxide, benzene and formaldehyde, among others. In addition, polynuclear aromatic hydrocarbons (PAHs), heterocyclic hydrocarbons, aromatic amines, volatile hydrocarbons, N-Nitrosamines, aldheydes, organic, inorganic and phenolic compounds were declared genotoxic environmental components [2]. In table 1 is shown the most important compounds of secondhand smoke.

Table 1. Most important compounds of secondhand smoke

Nicotine
Carbon Monoxide
Benzene
Formaldehyde
Polynuclear aromatic hydrocarbons (PAHs)
Heterocyclic hydrocarbons
Aromatic Amines
N-nitrosamines

Carcinogens in Secondhand Smoke

The International Agency for Research in Cancer (IARC), declared cadmium, beryllium, nickel, chromium, polonium-210, arsenic, ethylene oxide, benzene, vinyl chloride, 4-aminobiphenyl and 2-naphthylamine as Group I human carcinogens [3].

Table 2. IARC Group I carcinogens

cadmium
beryllium
nickel
chromium
polonium-210
arsenic
ethylene oxide
benzene
vinyl chloride
4-aminobiphenyl
2-naphthylamine

A special focus on certain types of Nitrosamines and its relation to cancer, have been developed. The first studies conducted to determine associations with cancer used experimental animals. Those studies showed a positive association between the component and cancer [4-5], establishing the carcinogenicity of most of the components. Important methodological aspects were considered to declare that a specific agent that causes cancer in animals also causes cancer in humans. For instance, qualitative and quantitative aspects were assessed, like: experimental conditions, consistency of results, interaction between agents, tumor development, dose, route and time of exposure, dose-response relationship, target organ, among others [6]

The causality principle that was evoked for this purpose was biological plausibility, because considering the absence of proper studies in humans, when the association with animals was sufficiently proven, a regard must be taken in the exposure to humans.

Quantification of Secondhand Smoke Exposure

Exposure to secondhand smoke can take place in any of the environments where people spend time. A useful conceptual framework for considering exposure to secondhand smoke is offered by the microenvironmental model that describes personal exposure to secondhand smoke as the weighted sum of the concentrations of secondhand smoke in the microenvironments where time is spent and the weights supplied by the time spent in each. A microenvironment is a space, for example a room in a dwelling or an office area, with a relatively uniform concentration of secondhand smoke during the time that is spent in that particular microenvironment. For research purposes and for considering health risks, personal exposure is the most relevant measure for evaluating and projecting risk.

Within the framework of the microenvironmental model, there are several useful indicators of exposure to secondhand smoke, ranging from surrogate indicators to direct measurements of exposure and of biomarkers that reflect dose. One useful surrogate, and the only indicator available for many countries, is the prevalence rate of smoking among men and women. It provides at least a measure of likelihood of exposure [7]

The indirect measures are generally obtained by questionnaires. These measures include self-reported exposure and descriptions of the source of secondhand smoke, in relevant microenvironments, most often the home and workplace. Self-reported exposure to secondhand smoke in the home and the workplace is a useful indicator of being exposed, although questionnaire-based reports of intensity of exposure are of uncertain validity [7].

Regarding direct measures of the components of second hand smoke, two main approaches have been used to quantify indoor concentrations and exposure: direct methods, using personal monitors and/or measuring biological markers, and indirect measurement methods, using ETS markers and/or mass balance modeling. Biological markers, which are components of ETS or their metabolites found in human physiological fluids, are the best direct means of assessing ETS exposure. Several components of ETS have been studied as markers for exposure to SHS. The ability to quantify exposure objectively is an important step in linking exposure to relative risk of adverse outcomes. Cotinine, a metabolite of nicotine, is a biological marker used in many epidemiological studies. Cotinine levels differ between smokers and ETS-exposed non-smokers by 2 – 3 orders of magnitude. Nicotine in hair is the most used biomarker in population-based studies. It has been shown to better correlate with exposure than cotinine [8] (Table 3).

Table 3. Quantification of secondhand smoke

Indirect measures	Direct measures
Questionnaires: Self-reported exposure	Personal monitors
Questionnaires: Description of the source of secondhand smoke	Biological markers (e.g. Cotinine)
	Environmental tobacco smoke
	Mass Balance modeling

3. Exposure to Secondhand Smoke

Second-hand smoke is still one of the most common indoor pollutants. Because of the time spent at the workplace and in the residential setting, these environments are considered to be the most important places of involuntary tobacco smoke exposure. Preventing from secondhand exposure in workplace and households has a remarkable impact in preventing form disease and disability.

Exposure to environmental tobacco smoke is an important problems worldwide. A retrospective study of burden of disease from exposure to second-hand smoke in 192 countries estimated that as many as 40% of children, 35% of women, and 33% of men are regularly exposed to secondhand smoke indoors [9]. The highest proportions exposed were estimated in Eastern Europe, the western Pacific, and the Southeast Asia, closely related to active smoking rates where no robust and extensive smoke-free indoor policies exist. Factors contributing to the differences in doses of indoor exposure to tobacco smoke between developing and developed countries include intensity of tobacco smoking, natural ventilation allowed by a comfortable climate, crowding at home and enforcement of legal protection from exposure to second-hand smoke in indoor workplaces and public places.

Intensity of secondhand smoke has relation with the environmental conditions under which people live and work. For demonstrate this the California Environmental Protection Agency established that children can roughly be divided into three exposure groups: children who have little or no exposure to ETS, children with smoking parents or guardians who take some measures to limit their child's exposure, and children highly exposed to ETS through smoking parents, guardians, or peer groups. Likewise, adults generally have virtually no exposure, experience regular but limited exposure in a public place, or experience substantial exposure through extensive contact with smokers. For individuals living in non-smoking homes and having only brief encounters with ETS, their average 24-hour exposure concentrations are low, and are estimated to be less than 0.01 μg/m3. For those living in homes with indoor smokers and experiencing in-vehicle exposures, their average exposure concentration to which they are exposed to over 24-hours can range up to 7.4 μg/m3. Such exposures are especially of concern for developing young children because they are likely to recur daily and may result in serious health consequences. The primary, and often the only exposure for individuals that do not spend time near smokers, occurs outdoors in locations over which the individual typically has little control. For non-smokers whose work or other activities bring them into contact with outdoor smokers regularly, 100% of their exposure can be attributable to proximity to outdoor smoking **[8].** These conclusions remark the importance of including outdoor areas in smoke-free legislation to safeguard health of the general population.

4. Effects of Secondhand Smoke on Health

Reports of the harmful effects of second-hand smoke exist since 1928, when it was proven to be related to lung tumors [10]. Scientific evidence about morbidity related to second-hand smoke has accumulated from many studies developed in different regions of the world and has established that exposure to tobacco smoke causes death, disease and disability. Regarding specific diseases, secondhand smoke has been recognized as a risk factor for chronic respiratory symptoms and diseases, low lung function, coronary heart disease and cancer [11]. In figure 1 is shown the major effects of secondhand smoke on health and in Table 4 is shown the evidence related with this effects.

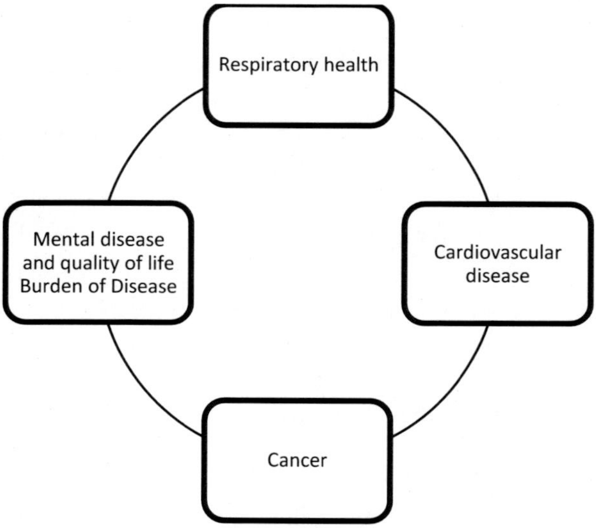

Figure 1. Major effects of secondhand smoke on human health.

Table 4. Secondhand smoke effects on human health

Health	Reference
Respiratory disease: wheeze; chronic cough; Phlegm; throat problems; decreased lung function; asthma; chronic bronchitis; chronic rhinosinusitis; among others.	[12-19]
Cardiovascular disease: mainly myocardial infarction, cardiac dysfunction	[18;20-26]
Cancer: Lung cancer; leukemia; breast cancer; upper aerodigestive tract carcinomas; nasal cancer; among others.	[28-36]
Mental disease and quality of life: Sleeping disorders; mental illness; social functioning; among others.	[37-38]

Respiratory and Cardiovascular Disease

Respecting respiratory symptoms, cross-sectional and longitudinal studies reported that exposure to secondhand smoke increases the risk of presenting wheeze, chronic cough, [12-15], phlegm and throat problems. There is also a higher risk of decreased lung function [12], asthma [16-18], chronic bronchitis [13,18], chronic rhinosinusitis [19], throat problems and frequent colds [15].

Concerning the association between SHS exposure and increased risks of cardiovascular morbidity and mortality, sufficient evidence is reported in epidemiological research to support this causal association [20-21] The risk of developing an acute cardiac syndrome or chronic lifetime coronary events is at least 30%. Similarly, reduction in the incidence of a myocardial infarction decreases by nearly 50% in the absence of SHS. [23]. It has been proven that passive smoking increases the risk of hypertension as well [18]. This disproportionately high risk of coronary heart disease, thought to be mediated through inflammation, platelet aggregation, and/or endothelial dysfunction. It has been widely investigated the relation between secondhand smoke exposure, measured objectively as cotinine, and recognized biomarkers of heart disease risk, namely C-reactive protein, homocysteine, fibrinogen, and white blood cell count. Workers exposed to low and high level of cotinine smokers appear to have disproportionately increased levels of two of these biomarkers of cardiovascular disease risk, fibrinogen and homocysteine [23-26].

Cancer

Evidence that exposure to passive smoking increases the risk of adults developing fatal diseases first emerged in 1981, related to lung cancer. Takeshi Hirayama published his epidemiological study demonstrating that secondhand smoke increased the risk of lung cancer in nonsmoking Japanese women married to men who smoked compared with non smoking women married to non-smoking men [28]. There were two other studies published at about the same time by Trichopoulos et. al [29] from Greece and Garfinkel [30] from the USA. While both studies showed an elevation in the point estimate of lung cancer risk associated with passive smoking, Garfinkel's study did not reach statistical significance. Since then, evidence about cancer related with second-hand smoke has accumulated from many studies done in these decades. Currently, there is a confirmed link between chronic secondhand smoke exposure and various types of cancer including lung cancer [31], leukemia [32-33], breast cancer [34], upper aerodigestive tract carcinomas [35] and nasal cancer [36].

Mental Disease and Quality of Life

Relating to mental health, several publications have appeared in recent years documenting the relation between secondhand smoking and sleeping disorders. One of them is a study performed in China with 219 children who were exposed regularly to SHS. Serum cotinine levels were used to measure exposure to tobacco smoke, and sleep patterns were assessed through parent reports. The results showed that exposure to SHS was associated with sleep problems, including longer sleep-onset delay, sleep-disordered breathing, parasomnias, daytime sleepiness, and overall sleep disturbance [37].

In regard to health-related quality of life (HRQOL), a cross-sectional study performed in Switzerland, showed that high SHS exposure at home had significantly lower scores on the physical functioning, role physical, bodily pain and social functioning. In the physical domains, there was a dose-response relationship, with higher daily SHS exposure resulting in greater reduction of health related quality of life, the association being stronger when women were exposed at home [38].

Burden of Disease

With respect to general population's health, a retrospective study held on 192 countries estimated the worldwide burden of disease attributable to second-hand smoke in 2004, measured as deaths and disability-adjusted life-years (DALYs) lost for non-smokers children and adults. This study concluded that worldwide second-hand smoke was estimated to have caused 603000 premature deaths and the loss of 10·9 million DALYs in 2004, corresponding to 1·0% of all deaths and 0·7% of the worldwide burden of disease in DALYs in this year [12]. The largest number of estimated deaths attributable to second-hand smoke exposure in adults was caused by ischemic heart disease, followed by lower respiratory infections in children, and asthma in adults. In assessment of burden of disease in terms of DALYs lost because of exposure to second-hand smoke, most DALYs lost were from lower respiratory infections, followed by those from ischemic heart disease and then from asthma in adults. Almost half of the total burden attributable to exposure to second-hand smoke was in Southeast Asia and the Western Pacific, with a high burden of disease also estimated in Eastern Europe, Africa, and Eastern Mediterranean region. In non-smokers, there are clear inequalities in the burden of disease from second-hand smoke according to sex and age. Women have the greatest burden of deaths of the total attributable to second-hand smoke, whereas children are most affected in terms of DALYs.

Effects on Health of Smoke- Free Legislation

Evaluations of the effects and effectiveness of workplace smoking policies have used a wide variety of study designs and measurements of exposure to secondhand tobacco smoke and tobacco use behaviors. In a review of the effectiveness of smoke-free legislation in enclosed places, the exposure to second-hand smoke in high-risk settings (such as bars and restaurants) was typically reduced by about 90%, and the exposure of adult non-smokers in the general population to second-hand smoke cut by as much as 60% [39]. Studies of the effects of smoke-free laws have drawn attention to the importance of this type of legislation in preventing disease and disability. For instance, smoke-free laws banning smoking in indoor workplaces rapidly reduce the prevalence of respiratory and sensory irritation symptoms in hospitality workers [40,41]. As an example, a mean reduction in the rate of hospitalizations for childhood asthma of 18% per year was observed in Scotland, following a comprehensive ban in 2006, compared with a mean annual rise of 5% before the implementation of the law [42].

Another well studied consequence is the decrease in numbers of acute coronary events. [43,44]. A meta-analysis considering 17 studies shows a pooled estimated reduction of 10% in the incidence of acute coronary events after the implementation of smoke-free legislation in six countries [45] Another meta-analysis of 12 studies published in Circulation founded that communities experience a 15% drop in heart attacks during the first year of smokefree legislation. [43].

5. Comprehensive Smoke-Free Legislation

According to the World Health Organization Framework Convention on Tobacco Control, countries that signed must implement comprehensive smoke-free legislation that prohibit tobacco smoking in all indoor public places, nevertheless, there are countries that still do not legislate, allowing to smoke in all areas, or have partial ban legislations, consenting to have mixed-areas inside a venue.

The WHO Framework Convention on Tobacco Control [46] is an international public health treaty adopted by the World Health Assembly on 21 May 2003. It entered into force on 27 February 2005. It had been signed by 168 countries and is legally binding in 179 ratifying countries. The FCTC is a global agreement that seeks "to protect present and future generations from the devastating health, social, environmental and economic consequences of tobacco consumption and exposure to tobacco smoke" [46]. The treaty represents a paradigm shift in developing a regulatory strategy to address tobacco use, asserting the importance of demand reduction strategies as well as supply issues.

One of the measures aiming to reduce the demand for tobacco is the protection from exposure to tobacco smoke, contained in Article 8 of the FCTC. This article addresses the adoption and implementation of effective measures to provide protection from exposure to second-hand tobacco smoke in indoor workplaces, public transport, indoor public places and, as appropriate, other public places. The prohibition of smoking in certain places seeks to provide a clean, safe and healthy environment for the public and to safeguard them from the harmful effects of second-hand smoke. [46].

Overall, nearly 80% of countries adopted or strengthened tobacco control legislation after ratifying the WHO FCTC up to 2014, but one third of the countries have still not put in place legislative measures in line with protecting policies from the noxious effects of tobacco smoking.

Regarding the protection from exposure to secondhand smoke, article 8 concerning smoke-free legislation has the highest implementation rate, with an average of 84%, across the 130 countries analyzed in the Global Progress report of 2014. If, however, only complete smoking bans are in the count, the average implementation rate declines up to 61%. [47]

Many countries reported that they have introduced legislation requiring a complete ban on smoking in various public places. A notable trend is the extension of smoking bans to public outdoor areas and to the use of novel products such as electronic cigarettes. Regarding this devices, the report on “Electronic nicotine delivery systems” by the WHO states that should the use of e-cigarettes indoors in public and work places should be restricted, bacause evidence suggests that exhaled e-cigarette aerosol increases the background air level of some toxicants, nicotine and particles. The report also refers to regulation required in advertising, promotion and sponsorship, to ensure that it does not target youth and non-smokers or people who do not currently use nicotine and will not serve as a gateway to nicotine addiction [47]

In the majority of cases smoke-free regulations are undertaken through national legislation, in other cases by administrative and executive orders or a combination of the two. A minor part of the countries are still using voluntary agreements to ensure protection from exposure to tobacco smoke.

The 2014 Global Progress Report on implementation of the WHO FCTC shows that apart from airplanes and ground public transport, health-care facilities, schools, government buildings and universities are the settings most frequently covered by a complete ban on tobacco smoking, while private workplaces, pubs and bars and especially private vehicles are the least covered. Several Parties also acted at subnational level. For example, in most of the states in Australia smoking is banned in private motor vehicles when minors are present, and in some cases in vehicles being used for business if anyone else is in the vehicle. Numerous municipalities in Canada have adopted bylaws or policies to prohibit smoking in public places such as patios, playgrounds and parks. In China, 12 cities have introduced local laws creating smoke-free environments, as well as putting in place mechanisms for their enforcement and imposition of penalties. Local authorities in New Zealand have continued to extend smoke-free areas within their jurisdictions. Smoke-free parks, playgrounds sports grounds, etc. are common. Recently some local council has begun to extend smoke-free areas to include selected streets/areas of town and bus shelters.

An education campaign leading to the implementation of newly adopted legislation will increase the likelihood of smooth implementation and high levels of voluntary compliance. Only fourteen countries have undertaken such campaigns, which is far below the expected rates.

In relation to the mechanism for enforcement of measures to protect their populations from exposure to tobacco smoke, the extension of smoking bans has required shared responsibility between various actors, most often local government, health, food and work safety authorities, and the police, to enforce smoke-free regulations in both indoor and outdoor areas.

Concluding Remarks and Perspectives

There is substantial evidence of the harmful consequences on health of secondhand exposure regarding it carcinogenic and non-carcinogenic effects.

Studies of the effects of smoke-free laws have drawn attention to the importance of second-hand smoke as a preventable cause of disease and disability.

Effective measures to provide protection from exposure to second-hand tobacco smoke in indoor workplaces, public transport, indoor public places and, as appropriate, other public places should be implemented in every country worldwide and this goal still has not been achieved.

REFERENCES

[1] Peto R, Lopez AD, Boreham J, et al. Mortality from tobacco in developed countries: indirect estimation from national vital statistics. *Lancet* 1992;339:1268–1278.

[2] Eriksen M, Chaloupka F. The economic impact of clean indoor air laws. *Cancer J Clinicians* 2007;57:367–78.

[3] WHO Report on the Global Tobacco Epidemic, 2008: the MPOWER package. Geneva: *World Health Organization,* 2008.

[4] Henderson AJ. The effects of tobacco smoke exposure on respiratory health in school-aged children. *Paediatr Respir Rev.* 2008; 9: 21–27.

[5] Hoffmann D, Hoffmann I, El-Bayoumy K. The less harmful cigarette: A controversial issue. A tribute to Ernst L Wynder. *Chem Res Toxicol* 2001; 14: 767-790.

[6] IARC. Monographs on the Evaluation of Carcinogenic Risks to Humans. Vol 38, Tobacco Smoking, Lyon, *IARC Press.*

[7] Wilbourne J, Haroun L, Heseltine E, Kaldor J, Partensky C, Vainio H. Response of experimental animals to human canrcinogens: an analysis based upon the IARC Monographs Programme. *Carcinogenesis* 1986;7:1853-1863.

[8] Tomatis L, Aitio A, Wilbourne J, Shuker L. Human carcinogens so far identified. *Jpn J Cancer Res* 1989;80:795-807.

[9] IARC. Monographs on the Evaluation of Carcinogenic Risks to Humans. Vol 89, Smokeless Tobacco and Some Tobacco-specific N-Nitrosamines, Lyon, *IARC Press.*

[10] IARC. Monographs on the Evaluation of Carcinogenic Risks to Humans. Vol 83, Tobacco Smoke and Involuntary Smoking, Lyon, *IARC Press.*

[11] California Environmental Protection Agency: Air Resources Board. Proposed Identification of Environmental Tobacco Smoke as a Toxic

Air Contaminant. *Sacramento: California Environmental Protection Agency*, 2005.

[12] Öberg M, et al. Worldwide burden of disease from exposure to second-hand smoke: a retrospective analysis of data from 192 countries. *The Lancet.* 2010;377:139 - 146

[13] Schönherr E. Contribution to the statistical and clinical features of lung tumours (in German). *Z Krebsforsch* 1928; 27: 436–50

[14] Hackshaw AK, Law MR, Wald NJ . The accumulated evidence on lung cancer and environmental tobacco smoke. *BMJ* 1997;315:980- 988.

[15] Hersoug LG, Husemoen LL, Sigsgaard T, Madsen F, Linneberg A. Indoor exposure to environmental cigarette smoke, but not other inhaled particulates associates with respiratory symptoms and diminished lung function in adults. *Respirology.* 2010; 15:993-1000.

[16] Radon K, Büsching K, Heinrich J, et al. Passive Smoking Exposure: A Risk Factor For Chronic Bronchitis And Asthma In Adults? *Chest.* 2002;122:1086-1090.

[17] Pirastu R, Bellu C, Greco P, Pelosi U, Pistelli R, Accetta G, Bigger A. Indoor exposure to environmental tobacco smoke and dampness: Respiratory symptoms in Sardinian children—DRIAS study. *Environmental Research.* 2009;109:59-65.

[18] Ho SY, Lam TH, Chung SF, Lam TP. Cross-Sectional and Prospective Associations Between Passive Smoking and Respiratory Symptoms at the Workplace. *Annals of Epidemiology.* 2007;17:126–131.

[19] Leuenberger P, Schwartz J, Ackermann-Liebrich U, et al. Swiss Study on Air Pollution and Lung Diseases in Adults, SAPALDIA Team, Passive smoking exposure in adults and chronic respiratory symptoms (SAPALDIA Study). *Am J Respir Crit Care Med* 1994;150:1222- 1228.

[20] Thorn J, Brisman J, Torén K. Adult-onset asthma is associated with self-reported mold or environmental tobacco smoke exposures in the home. *Allergy* 2001;56:287- 292.

[21] Vozoris N, Lougheed MD. Second-hand smoke exposure in Canada: Prevalence, risk factors, and association with respiratory and cardiovascular diseases. *Canadian Respiratory Journal: Journal of the Canadian Thoracic Society* 2008;15:263-269.

[22] Tammemagi C, Davis RM, S. Benninger M, Holm AL, Krajenta R. Secondhand Smoke as a Potential Cause of Chronic Rhinosinusitis: A Case-Control Study. *Arch Otolaryngol Head Neck Surg.* 2010;136(4):327-334.

[23] Law MR, Morris JK, Wald NJ. Environmental tobacco smoke exposure and ischaemic heart disease: an evaluation of the evidence. *BMJ* 1997; 315: 973- 980.
[24] Carey IM, Cook DG, Strachan DP. The effects of environmental tobacco smoke exposure on lung function in a longitudinal study of British adults. *Epidemiology* 1999;10(3):319- 326.
[25] Faught BE, Flouris AD, Cairney J. Epidemiological evidence associating secondhand smoke exposure with cardiovascular disease. *Inflamm Allergy Drug Targets.* 2009;8:321-327.
[26] Venn A, Britton J.Exposure to Secondhand Smoke and Biomarkers of Cardiovascular Disease Risk in Never-Smoking Adults. *Circulation.* 2007;115:990-995,
[27] Clark JD 3rd, Wilkinson JD, LeBlanc WG, Dietz NA, Arheart KL, Fleming LE, Lee DJ. Inflammatory markers and secondhand tobacco smoke exposure among U.S. workers. *Am J Ind Med.* 2008;51:626-32.
[28] Hirayama T. Non-smoking wives of heavy smokers have a higher risk of lung cancer: a study from Japan. *BMJ* 1981;282:183-185.
[29] Trichopoulos D, Kalandidi A, Sparros L, MacMahon B. Lung cancer and passive smoking. *Int J Cancer* 1981;27:1-4.
[30] Garfinkel L. Time trends in lung cancer mortality among non-smokers and a note on passive smoking. *Journal of the National Cancer Institute.* 1981, 66: 1061–1066.
[31] Besaratinia A, Pfeifer GP. Second-hand smoke and human lung cancer. *Lancet Oncol.* 2008;9:657–666.
[32] Chang JS. Parental smoking and childhood leukemia. *Methods Mol Biol.* 2009;472:103–137.
[33] Lee KM, Ward MH, Han S, Ahn HS, Kang HJ, Choi HS, Shin HY, Koo HH, SeoJJ, Choi JE, Ahn YO, Kang D. Paternal smoking, genetic polymorphisms in CYP1A1 and childhood leukemia risk. *Leuk Res.* 2009; 33: 250–258.
[34] Johnson KC. Accumulating evidence on passive and active smoking and breast cancer risk. *Int J Cancer* 2005;117: 619–628.
[35] Sturgis EM, Pytynia KB. After the smoke clears: environmental and occupational risks for carcinoma of the upper aerodigestive tract. *Cancer J.* 2005;11:96–103.
[36] Rushton L. Health impact of environmental tobacco smoke in the home. *Rev Environ Health.* 2004;19: 291–309.
[37] Yolton K1, Xu Y, Khoury J, Succop P, Lanphear B, Beebe DW, Owens J. Associations between secondhand smoke exposure and sleep patterns

in children. *Pediatrics*. 2010 Feb;125(2):e261-8. doi: 10.1542/peds.2009-0690. Epub 2010 Jan 18.

[38] Bridevaux P, Cornuz J, Gaspoz J, et al. Secondhand Smoke and Health-Related Quality of Life in Never Smokers: Results From the SAPALDIA Cohort Study 2. *Arch Intern Med.* 2007;167(22): 2516-2523.

[39] International Agency for Research on Cancer. Evaluating the effectiveness of smoke-free policies. Lyon: International Agency for Research on Cancer, 2009. http://www.iarc.fr/en/publications/pdfs-online/prev/handbook13/handbook13.pdf

[40] Hahn E, Lee K, Okoli C, Troutman A, et al. Smoke-free laws and indoor air pollution in Lexington and Louisville. *Louisville Med* 2005; 52:391–394, 409, 415.

[41] Bannon F, Devlin A, McElwee G, Gavin A. Greater gains from smoke-free legislation for non-smoking bar staff in Belfast. *The European Journal of Public Health.* 2009;19(6):638-643.

[42] Mackay DF, Haw S, Ayres JG, et al. Smoke-free legislation and hospitalizations for childhood asthma. *N Engl J Med* 2010;363:1139e45.

[43] Lightwood JM, Glantz SA. Declines in acute myocardial infarction after smoke-free laws and individual risk attributable to secondhand smoke. *Circulation* 2009; 120: 1373–79.

[44] Pell JP, Haw S, Cobbe S, et al. Smoke-free legislation and hospitalizations for acute coronary syndrome. *N Engl J Med* 2008; 359: 482–91.

[45] Mackay DF, Irfan MO, Haw S, et al. Meta-analysis of the effect of comprehensive smoke-free legislation on acute coronary events. *Heart.* 2010;96:1525-1530.

[46] WHO. Framework Convention on Tobacco Control. World Health Organisation, http://www.who.int/tobacco/framework/WHO_FCTC_english.pdf; 2003. p. 8.

[47] WHO, 2014 http://www.who.int/nmh/events/2014/backgrounder-e-cigarettes/en/

INDEX

#

A

B

C

D

E

F

J

K

L

M

N

O

P

Q

R

S

T

Z